SpringerBriefs in Energy

For further volumes:
http://www.springer.com/series/8903

Ibrahim Dincer • Anand S. Joshi

Solar Based Hydrogen Production Systems

 Springer

Ibrahim Dincer
Faculty of Engineering
and Applied Sciences
University of Ontario Institute
of Technology (UOIT)
Oshawa, ON, Canada

Anand S. Joshi
Faculty of Engineering
and Applied Sciences
University of Ontario Institute
of Technology (UOIT)
Oshawa, ON, Canada

ISSN 2191-5520   ISSN 2191-5539 (electronic)
ISBN 978-1-4614-7430-2   ISBN 978-1-4614-7431-9 (eBook)
DOI 10.1007/978-1-4614-7431-9
Springer New York Heidelberg Dordrecht London

Library of Congress Control Number: 2013936954

Printed on acid-free paper

Springer is part of Springer Science+Business Media (www.springer.com)

Preface

The book presents various hydrogen production systems in general, and solar hydrogen production systems in particular, and their energy and exergy analyses. This book is also concerned with various chemical processsess involved in hydrogen production. The primary energy sources, conventional and nonconventional, that are used to produce hydrogen are also studied in brief for analysis and comparison. The sustainability aspects of different methods are evaluated in terms of a sustainability index, and the environmental impact assessments of different methods are also evaluated in terms of the environmental impact reduction factor.

The most feasible and commercially practiced method of hydrogen production is water electrolysis. Many other methods, for example, thermolysis, thermochemical cycles, gasification, cracking, and reforming, are used to produce hydrogen. In addition to high-temperature solar hydrogen production, other methods such as photovoltaic-based hydrogen production, photo-electrolysis, and bio-photolysis are also covered here. Some other methods that are discussed in the book are hydrogen production through carbonization of bituminous coal, solar hydrogen production via biomass and hydrogen production by decarbonization of fossil fuels.

This book aims to provide a useful source for researchers, scientists, engineers, and practitioners working in the field of solar hydrogen production because it provides some quality analysis of solar thermal hydrogen production systems and solar photovoltaic-based hydrogen production systems.

Before closing, the authors acknowledge the financial support provided by the Natural Sciences and Engineering Research Council of Canada and Turkish Academy of Sciences and the material support by Dr. Dincer's PhD students, Ehsan Baniasadi, Tahir Ratlamwala, Ahmet Ozbilen and Ahmet Yilanci.

Oshawa, ON, Canada   Ibrahim Dincer
Oshawa, ON, Canada   Anand S. Joshi

Contents

Chapter 1
Hydrogen

1.1 Introduction

Hydrogen, which is the first element in the periodic table of elements, having the atomic number 1, is a non-metal, colorless, odorless, highly reactive, self-burning characteristic gas that has all the qualities to be a fuel for our automobiles and a future energy source. In contrast to conventional energy sources (coal, oil, natural gas), it is carbon free and hence environmentally friendly. The countdown for conventional sources of energy has already begun as they are depleting fast. Therefore, hydrogen is a perfect candidate to fulfill the energy needs of humans for the future. In addition to the aforementioned qualities, hydrogen is quite challenging as an energy source or fuel because of its availability. Although hydrogen is naturally present on the Earth in the combined state in both organic and inorganic compounds, for example, as water and hydrocarbons, it is scarcely present in the free and molecular state. Therefore, elemental hydrogen is artificially produced, and hence its safe and environmentally benign production is most important. When considering environmentally friendly hydrogen production, the obvious choice for the input energy is renewable energy, mainly solar energy.

1.2 Hydrogen as a Potential Fuel

Historically, human beings have needed energy for their survival. In the beginning we learned the power of fire and then its applications, for example, in heating and for cooking food. After the invention of the steam engine in the Industrial Revolution in the 1860s, we learned the power of fossil fuels, and ever since then our energy needs have been increasing. The only changing parameter in the foregoing scenario is the type of fuel we are using.

With advances in fuel technology, we started using the best fuel available, starting from coal, then on to oil and natural gas. All these fuels are known as

I. Dincer and A.S. Joshi, *Solar Based Hydrogen Production Systems*,
SpringerBriefs in Energy, DOI 10.1007/978-1-4614-7431-9_1, © The Author(s) 2013

fossil fuels, and our automobiles, industries, and all the energy-intensive uses are dependent on these alone. In other words, we became dependent on fossil fuels and harvested coal, oil, and gas more and more intensively to match our energy demand, to reduce the cost of fuel, to increase the production of the goods we needed, to maintain the high standards of living in our society, and so on. However, the intensive mass exploration of fossil fuels caused two main side effects, similar to any prescription drug, namely, (1) the fast depletion of fossil fuel reserves and (2) the global problem of global warming, caused by environmental damage from greenhouse gas emissions as a by-product of the combustion of fossil fuels. The fast depletion of fossil fuel reserves is worrisome to many, including those in the power sector and various industries, whereas increasing greenhouse gas emissions is the worry of environmentalists and concerned others.

A question arises here: how to fulfill the ever-increasing energy demands without damaging our environment. One of the answers could be renewable energy sources such as solar, wind, geothermal, biomass, waves, tides, and hydropower. The advantages with renewable sources are that these are re-new-ables, which means never depleted and freely or cheaply and immensely available; these sources may be either free of greenhouse gas emissions (e.g., solar, wind, hydro) or less polluting (e.g., biomass). Some disadvantages with renewables include the intermittency of available energy and low potential energy sources. For example, in wind energy, there is no fixed wind pattern, which changes with time, or in the case of solar energy, the intensity of solar radiation falling on the Earth's surface also changes with time. Further, a large amount of extraterrestrial solar radiation (about 30 %) is absorbed or reflected by the atmosphere when the radiation passes through it. Despite all these odds, the potential of solar energy to fulfill human energy needs is immense. As mentioned earlier, only 70 % of the solar energy reaches Earth's surface after atmospheric absorption and reflection, which is about 3.9×10^{24} MJ incident solar energy per year [1]. According to Markandya and Wilkinson [1], this solar energy is 10,000 times greater than the current global energy consumption, and thus less than 1 % of solar energy would serve all the energy needs of humans.

To strengthen the foregoing discussion, the estimate of world fossil fuel production with the forecast of energy demand and supply is shown in Fig. 1.1. It is evident that coal and oil are being depleted very rapidly and will continue to do so in the foreseeable future. It can also be seen from this figure that world energy demand is increasing. To fulfill the world energy demands in the absence of fossil fuels, we need a very strong energy source that is sustainable and eco-friendly and which can competently power all the different energy-intensive sectors. The candidate is hydrogen, because it can be used as a fuel in internal combustion engines and can also be used to generate electricity. In other words, hydrogen can be used as both an energy carrier and a fuel for our vehicles. The primary energy sources, such as fossil fuels and solar, wind, and other renewable sources, are often used to generate electricity and heat as these are the two common forms of energy that can directly be utilized at the consumer end. In general, 25 % of the primary energy is utilized to generate electricity whereas 75 % is used to produce heat. The aforementioned primary energy sources must therefore be converted to these energy carriers needed

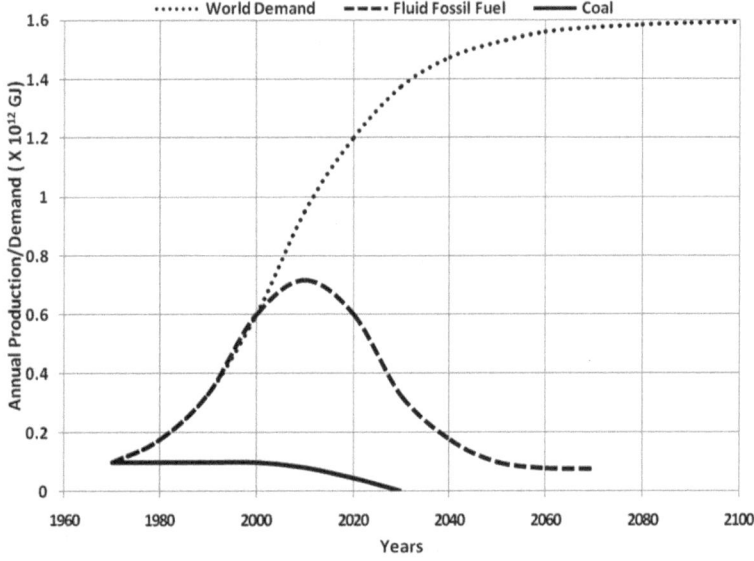

Fig. 1.1 Estimates of world fossil fuel production (Modified from Veziroglu and Barbir [2])

by the consumer. In contrast to fossil fuels, none of the renewable energy sources can be used directly as a fuel, for example, for air and land transportation, and therefore they must be used to produce fuel(s) as well as to generate electricity. Some suitable candidates for fuel are synthetic gasoline, synthetic natural gas (methane), ethanol, methanol, and hydrogen. Therefore, the fuel of choice for the post-fossil fuel era is expected to satisfy the following conditions [2–6]:

- It must be environmentally benign.
- It must be sustainable.
- Its source may be recycling water into hydrogen and oxygen and back to water via a fuel cell.
- It must be suitable for transportation.
- It must be versatile and able to be easily be converted to other energy forms at the user end.
- It must have high utilization efficiency.
- It must be safe to use.

1.3   Some Common Properties of Hydrogen

Hydrogen, under most conditions, is a colorless, odorless, tasteless, and nontoxic gas. It undergoes the phase change from gas to a liquid state at −423 °F (20.26 K or −252.89 °C). Hydrogen is the lightest of the elements in the periodic table, with a very low density per unit volume; it is approximately 14 times less dense than air.

It is very buoyant, as well as highly diffusive in the surrounding air; thus, it disperses in the atmosphere very quickly and can escape through very small spaces. These properties make hydrogen storage challenging for scientists and engineers because of the leakage problems. The same properties also enable hydrogen atoms to penetrate the molecular structure of some metals, making them "brittle," especially where physical stresses from extreme temperatures or elevated pressures exist. This phenomenon is known as "hydrogen embrittlement" [7]. The combustion properties of hydrogen make it suitable as a fuel and energy carrier. Its energy content per unit mass, which is also known as high heating value, is about three times that of gasoline. The high heating value can be defined as the amount of energy released when a fuel is burned completely. Hydrogen ignites across a wide range of concentrations, from about 4 % to 74 %, and only requires a small amount of energy to initiate ignition.

Hydrogen, when combined with oxygen in either combustion or an electrochemical conversion process, produces energy, and the by-product of the reaction is water vapor. Because of its highly reactive properties and characteristics, the immediate dangers associated with hydrogen leaks are fire, explosion, and asphyxiation. Hydrogen in presence of air is ignited immediately by coming in contact with sources of heat, open flames, electrical sparks, and static electric discharge. The flame characteristic of most hydrogen fires has a torch or jet originating at the point of hydrogen discharge. Hydrogen flames are difficult to detect as they burn hot, resulting in little radiant heat, and are thus invisible to the naked eye. This problem can be dangerous as the flames can be stumbled upon without warning. The risk of explosion of hydrogen becomes high because of its fuel properties, such as low density and high diffusivity, which cause explosions under pressurized conditions; however, the blast energy from its explosion is typically less than that resulting from the equivalent energy content from other fuels. Explosion occurs mainly by the formation of pressure waves followed by the formation of sound waves caused by sufficiently rapid combustion in a confined space/pressurized condition. A hydrogen explosion may cause significant damage. Furthermore, although hydrogen is odorless and nontoxic in nature, it can result in suffocation as it dilutes the concentration of oxygen in the air necessary for respiration to support human life. The risk of asphyxiation (a condition of being deprived of oxygen, e.g., by stoppage of breathing) thus cannot be avoided though it is quite uncommon.

The hazards of fire and explosion are primary hazards compared to asphyxiation as the amount of hydrogen that can cause an oxygen-deficient environment is well within the flammable range. Other than the foregoing fire hazards, liquid hydrogen presents an additional hazard because of its extremely cold (cryogenic) temperature. When talking about safety, however, hydrogen has numerous advantages over other fuels. The hydrogen fires can be rated safer than fires involving conventional fuels as the former has a very high flame speed and also it dissipates quickly compared to the latter, enabling hydrogen to burn fast, even in case of liquid hydrogen. Because of the fast-burning characteristics of hydrogen, other materials close to the flame are unlikely to catch fire themselves, thereby reducing the danger of toxic smoke emissions and prolonged burning.

1.4 Hydrogen Energy

The name hydrogen originated from two Greek words, "hydro" and "genes," meaning "water" and "generator." For more than half a century, hydrogen has been used in many chemical processes in industry and as a rocket propellant. As discussed in the previous section, during that time, hydrogen was produced using a very robust infrastructure. Then, it was stored, transported, and utilized with understanding and respect for its physical properties in the different energy-consuming sector. The primary energy used to produce hydrogen from water was fossil fuel at that time, but because of advances in technology and environmental issues, fossil fuels are being replaced by much cleaner and more sustainable renewable energy sources. Hydrogen is the main component of all other elements, consisting of one proton and one electron. Its highly reactive nature enables it to react easily with other elements to form various compounds, for example, water (H_2O) and hydrocarbon fuels (oil, natural gas, coal). Many scientists, environmentalists, and eco-lovers believe that hydrogen could be used as an energy carrier and transportation fuel. Its physical properties and characteristics define its suitability as an energy carrier or transportation fuel. The same properties and characteristics that make hydrogen an ideal energy carrier and fuel also make it safe to use at the consumer end. It has been seen in industry that hydrogen can be used safely in a wide range of applications and conditions by employing proper safety controls. Therefore, it is fair to say that hydrogen is "no more or no less dangerous" than other fuels.

Chapter 2
Hydrogen Production Methods

2.1 Introduction

As hydrogen appears to be a potential solution for a carbon-free society, its production plays a critical role in showing how well it fulfills the criteria of being environmentally benign and sustainable. Of course, hydrogen can be produced from a number of sources, such as water, hydrocarbon fuels, biomass, hydrogen sulfide, boron hydrides, and chemical elements with hydrogen. Because hydrogen is not available anywhere as a separate element, it needs to be separated from the aforementioned sources, for which energy is necessary to do this disassociation. The forms of energy that can drive a hydrogen production process can be classified in four categories: thermal, electrical, photonic, and biochemical energy. These kinds of energy can be obtained from primary energy (fossil, nuclear, and renewable) or from recovered energy through various paths. The literature is quite large and covers many options.

Many researchers have been involved in analyzing the different hydrogen production methods based on energy and exergy analysis. As mentioned by Muradov and Veziroglu [8], ammonia, being rich in hydrogen, can be used as a fuel directly [in internal combustion engines (ICE)] or via on-board decomposition to hydrogen and nitrogen (in ICE and fuel cells). Zamfirescu and Dincer [9] proposed a system that uses ammonia as the source of hydrogen. In their system, the heat recovered from an engine or fuel cell was used to extract hydrogen from ammonia. Yilanci et al. [10] have made a through and up-to-date review on the various hydrogen production systems and analyzed a solar–hydrogen–fuel cell hybrid energy system in terms of energy and exergy efficiencies for stationary applications in Denizli, Turkey. They reported that the overall energy efficiency values of the system vary between 0.88 % and 9.7 %, whereas minimum and maximum overall exergy efficiency values of the system are between 0.77 % and 9.3 %. Balta et al. [11] have analyzed a geothermal-based hydrogen production

I. Dincer and A.S. Joshi, *Solar Based Hydrogen Production Systems*,
SpringerBriefs in Energy, DOI 10.1007/978-1-4614-7431-9_2, © The Author(s) 2013

system for Iceland in terms of energy and exergy efficiency and reported that the efficiency varies with the geothermal inlet temperature. This process involves high-temperature steam electrolysis (HTSE) coupled with a geothermal source.

Abanades and Flamant [12] have studied the single-step thermal decomposition (pyrolysis) of methane without catalysts. The process coproduces hydrogen-rich gas and high-grade carbon black (CB) from concentrated solar energy and methane. It is an unconventional route for potentially cost-effective hydrogen production from solar energy without emitting carbon dioxide because solid carbon is sequestered. For the experiment with the 2-m-diameter concentrator, the thermochemical efficiency is in the range of 2–6 % for the maximum conversion (98 %), assuming that the mean temperature in the nozzle is 1,500 K. Liu et al. [13] investigated hydrogen production by integrating methanol steam reforming with a 5-kW solar reactor that can produce 150–300 °C at atmospheric pressure and obtained thermochemical efficiency of solar thermal energy converted into chemical energy in the range of 30–50 %.

Z'Graggen et al. [14] analyzed hydrogen production by steam-gasification of petroleum coke using concentrated solar power and reported a solar energy conversion efficiency of 17 %. Charvin et al. [15] made a process analysis of ZnO/Zn, Fe_3O_4/FeO, and Fe_2O_3/Fe_3O_4 thermochemical cycles and found these to be potentially high-efficiency, large-scale, and environmentally attractive routes to produce hydrogen by concentrated solar thermal energy that operates at a temperature up to 2,000 K. The real energy efficiency of these cycles was reported as 25.2 %, 28.4 %, and 22.6 %, respectively. Falco et al. [16] reported that the application of hydrogen-selective membranes (for example, a Pd/Ag membrane) in steam reforming plants may play an important role in converting natural gas or heavy hydrocarbons into hydrogen in a very efficient way, and by providing the reaction heat by sources such as solar-heated molten salts or a fluid heated in a nuclear reactor may further increase the overall energy efficiency of the system and pave the way for producing large amounts of hydrogen with minimum environmental impact.

Ni et al. [17], have conducted energy and exergy analyses of the thermodynamic-electrochemical characteristics of hydrogen production by a PEM electrolyzer plant and found that the energy and exergy efficiencies of the system are same and influenced by the operating temperature, current density, and the thickness of the polymer electrolyte membrane (PEM). With an increase in current density from 2,000 to 10,000 A/m^2, an operating temperature of 353 K, and a PEM electrolyte thickness of 100 μm, the efficiency decreases from 0.64 to 0.58. They also claimed that with an increase in the thickness of the PEM electrolyte and the operating temperature, the efficiency of the plant is reduced. For the three different PEM electrolyte thicknesses, that is, 50, 100, and 200 μm (and at 10,000 A/m^2 current density), the energy efficiency is 0.6, 0.58, and 0.56 respectively. For three different operating temperatures (300, 323, and 353 K) the energy efficiency is 0.55, 0.57, and 0.58 at a current density of 10,000 A/m^2. For higher current densities the difference in efficiency is more evident than for lower current densities.

Zedtwitz et al. [18] have produced hydrogen via solar thermal decarbonization of fossil fuels using three different routes and reported an exergy efficiency of 32 %

for solar decomposition of natural gas, 46 % for solar steam reforming of natural gas, and 46 % for solar steam gasification of coal. Although the exergy efficiency of the first route is less as compared to the latter two, it is a zero carbon dioxide emission method of producing hydrogen.

2.2 Classification of Hydrogen Production Methods

Hydrogen can be produced by both renewable and nonrenewable sources of energy. The former has the advantage of being environmentally friendly whereas the latter has either carbon dioxide or some other form of carbon residue in the end product other than hydrogen. Hydrogen production using conventional sources, that is, coal, oil, and natural gas, is in practice these days, and research is ongoing to minimize the environmental damage caused by greenhouse gas emissions. One method by which greenhouse gases can be minimized is by using solar or some other form of renewable energy source as the primary energy requirement for the hydrogen production chemical reaction. Therefore, it is important to understand the renewable energy sources first and then how these energy sources can be used for hydrogen production. Dincer [19] has summarized various green hydrogen production methods that use renewable energy sources (Table 2.1).

Careful reading of Table 2.1 shows that the primary energy required for the chemical reactions is generally electrical and thermal energy. The materials or chemicals used to generate hydrogen are principally water and fossil fuels. Organic biomass and inorganic compounds such as hydrogen sulfide are also used to produce hydrogen. Therefore, it is important to identify the sources of energy that can be used to fulfill the primary energy demands for environmentally benign hydrogen production.

The energy conversion from energy sources to process energy is equally important, as summarized by Dincer [19] in Table 2.2. It is important to see that electricity may be produced by all the renewable energy sources. High-grade thermal energy can be produced by concentrated solar energy, biomass and recovery gas from landfills, etc., and low-grade thermal energy can be produced geothermally.

Taking the foregoing discussion further, this section considers hydrogen production using renewable and sustainable energy resources, for example, solar, wind, and geothermal. Hydrogen production mainly involves thermal and electrical energy as the input energy; therefore, different renewable sources are used to provide input energy. Because most of the renewable sources are used to produce electricity first and the electricity is then further utilized to produce hydrogen, for example, in an electrolyzer unit, different electricity production methods are also discussed briefly here. Some renewable sources, for example, geothermal, can also be used to produce heat that can be used in thermochemical and hybrid cycles for hydrogen production. Discussion of different modes of hydrogen production, that is, via electricity and via thermal, appears in this chapter as necessary.

Table 2.1 Classification of green hydrogen production methods

Primary energy	Hydrogen production method	Material resources	Brief description
Electrical energy	Electrolysis	Water	Water decomposition into O_2 and H_2 by passing a direct current which drives electrochemical reactions
	Plasma arc decomposition	Natural gas	Clean natural gas (methane) is passed through an electrically produced plasma arc to generate hydrogen and carbon soot
Thermal energy	Thermolysis	Water	Steam is brought to temperatures of over 2,500 K at which water molecule decomposes thermally
	Thermo-catalysis	H_2S cracking	Hydrogen sulfide	H_2S extracted from sea or derived from other industrial processes is cracked thermo-catalytically
		Biomass conversion	Biomass	Thermo-catalytic biomass conversion to hydrogen
	Thermochemical processes	Water splitting	Water	Chemical reactions (including redox reactions or not) are conducted cyclically with overall result of water molecule splitting
		Gasification	Biomass	Biomass converted to syngas; H_2 extracted
		Reforming	Biofuels	Liquid biofuels converted to hydrogen
		H_2S splitting	Hydrogen sulfide	Cyclical reactions to split the hydrogen sulfide molecule
Photonic energy	PV electrolysis	Water	PV panels generate electricity to drive electrolyzer
	Photo-catalysis	Water	Complex homogeneous catalysts or molecular devices with photo-initiated electron collection are used to generate hydrogen from water
	Photo-electrochemical method	Water	A hybrid cell is used to generate photovoltaic electricity, which drives the water electrolysis process
	Bio-photolysis	Water	Biological systems based on cyanobacteria are used to generate hydrogen in a controlled manner

Biochemical energy	Dark fermentation	Biomass	Anaerobic fermentation in the absence of light
Electrical + thermal	Enzymatic	Water	Uses polysaccharides to generate the required energy
	High-temperature electrolysis	Water	Uses a thermal source and electrical power to split water in solid oxide electrolyte cells
	Hybrid thermochemical cycles	Water	Use thermal energy and electricity to drive chemical reactions cyclically with the overall result of water splitting
	Thermo-catalytic fossil fuel cracking	Fossil fuels	A thermo-catalytic process is used to crack fossil hydrocarbons to H_2 and CO_2, whereas CO_2 is separated/sequestrated for the process to become green
	Coal gasification	Water	Coal is converted to syngas, then H_2 extracted and CO_2 separated/sequestrated (electric power spent)
	Fossil fuels reforming	Fossil fuels	Fossil hydrocarbons are converted to H_2 with CO_2 capture and sequestration (electric power spent)
Electrical + photonic	Photo-electrolysis	Water	Photo-electrodes + external source of electricity
Biochemical + thermal	Thermophilic digestion	Biomass	Uses biomass digestion assisted by thermal energy for heating at low-grade temperature
Photonic + biochemical	Bio-photolysis	Biomass, water	Uses bacteria and microbes to photo-generate hydrogen
	Photo-fermentation	Biomass	The fermentation process is facilitated by light exposure
	Artificial photosynthesis	Biomass, water	Chemically engineered molecules and associated systems to mimic photosynthesis and generate H_2

Modified from Dincer [19]

Table 2.2 Production methods and energy conversion paths to produce "green hydrogen"

Hydrogen production method	Green energy source	Conversion path
Electrolysis (green energy generates electricity for water electrolysis) or *plasma arc decomposition* (green energy generates electricity for plasma arc decomposition of natural gas)	Solar	PV power plant or concentrated solar power (CSP) to generate electricity
	Geothermal	Power plant [organic Rankine cycle (ORC), flash cycle, etc.]
	Biomass	Biomass power plant, internal combustion engines, fuel-cell plants
	Wind	Wind power plants (grid-connected or autonomous)
	Ocean heat	OTEC (ocean thermal energy conversion) plants
	Other renewable	Tides, ocean currents, and wave energy converted into electricity
	Nuclear	Nuclear power plants
	Recovery	Landfill gas combusted in diesel generators
		Industrial/other heat recovery used to drive ORC or other heat engines
		Incineration with pollutant capture drives Rankine power plant
Thermolysis		
Thermo-catalysis	Solar	Concentrated solar heat used to generate ultrahigh-temperature steam
	Solar	Concentrated solar heat used to drive the process at high temperature
	Biomass	Low-grade biomass combustion generates the process heat
	Recovery	Landfill gas combustion, high-temperature industrial heat recovery
Biomass conversion	Solar	Concentrated solar heat at high temperature drives the process
	Biomass	Auto-thermal process: reaction heat comes from biomass combustion
Thermochemical processes		
Water splitting	Solar	Concentrated solar radiation generates high-temperature heat
	Geothermal	Geothermal-generated electricity to drive high-temperature heat pumps
	Biomass	Dried biomass is combusted to generate high-temperature heat
	Nuclear	Nuclear electric power used to drive high-temperature heat pumps
	Recovery	Landfill gas combustion
Gasification	Solar	Concentrated solar heat at high temperature drives the process
	Biomass	Auto-thermal process: reaction heat comes from biomass combustion
Fuel reforming	Solar	Concentrated solar heat at high temperature drives the process
	Biofuels	Auto-thermal process: reaction heat comes from biomass combustion
H_2S splitting	Solar	Concentrated solar heat used to drive the process at high temperature
	Geothermal	High-temperature geothermal heat at ~200 °C drives the process

Method	Energy source	Description
	Biomass	Low-grade biomass combustion generates the process heat
	Recovery	Landfill gas combustion, high-temperature industrial heat recovery
PV electrolysis	Solar	Solar radiation generates electricity through PV panels
Photo-catalysis	Solar	UV and upper spectrum visible solar radiation drives the process
Photo-electrochemical	Solar	All solar spectrum used by photo-electrochemical cell
Bio-photolysis	Solar	All solar spectrum can be used
Dark fermentation	Biomass	Biogas reactors are used for dark fermentation to generate hydrogen
Enzymatic	Biomass	Polysaccharides are manipulated by special enzymes to extract hydrogen
High-temperature electrolysis	Solar	Concentrated solar power generates high-temperature heat and electricity
	Geothermal	Geothermal electricity coupled to high-temperature heat pumps
	Biomass	Biomass combustion generates power and high-temperature heat
	Nuclear	Nuclear power used to generate electricity and high-temperature heat
	Recovery	Recovered energy generates electricity and high-temperature heat
Hybrid thermochemical cycles	Solar	Concentrated solar power generates high-temperature heat and electricity
	Geothermal	Geothermal electricity coupled to high-temperature heat pumps
	Biomass	Biomass combustion generates power and high-temperature heat
	Nuclear	Nuclear power used to generate electricity and high-temperature heat
	Recovery	Recovered energy generates electricity and high-temperature heat
Photo-electrolysis	Solar	PV or CSP + electrolysis bath with photo-electrodes
Thermophilic digestion	Biomass + other	Biomass energy drives the process; heat recovery or solar provides heat
Bio-photolysis	Biomass + solar	Biomass + photonic energy drive the process
Photo-fermentation	Biomass + solar	Biomass + photonic energy drive the process
Artificial photosynthesis	Solar	Solar energy drives the hydrogen generation process directly

A classification of solar hydrogen production systems based on energy input [that is, sunlight (photo) and solar thermal, and type of chemical reactants, for example, H_2O, natural gas, oil, coal] and for the different hydrogen production processes involved, for example, electrolysis, reforming, gasification, and cracking, is also presented. Thermochemical cycles, such as the hybrid-sulfur cycle, metal oxide-based cycle, and electrolysis of water are the most promising processes for environmentally benign future hydrogen production. The concept of sustainable and environmentally benign hydrogen production by artificial photosynthesis is also discussed. For a case study, sustainability of a solar hydrogen system through exergy efficiency and sustainability index is investigated. The various processes associated with solar hydrogen production in terms of exergy efficiency and sustainability index are also compared.

2.3   Renewables for Hydrogen Production

In this section, hydrogen production via renewable sources is discussed. As already mentioned, thermal and electrical energy are the input energy sources; therefore, in this section a brief discussion about these is included. Electricity can be produced by various renewable resources, such as solar, wind, geothermal, tidal, wave, ocean thermal, hydro, and biomass. Generally, with these technologies, the electricity produced is supplied to the grid, but with some technologies, for example, solar photovoltaic, the electricity can also be supplied to small standalone systems. Renewable sources of energy are known as eco-friendly and sustainable energy resources, in contrast to fossil fuels (coal, oil, natural gas) that produce greenhouse gases such as carbon dioxide, which are responsible for global warming on this planet Earth. Moreover, fossil fuels are finite sources and they are depleting fast. Some established renewable technologies for electricity and thermal energy production are discussed briefly here. Also, the various processes involved in hydrogen production, such as electrolysis, thermolysis, photo-electrolysis, and photosynthesis, are discussed in connection with renewable energy sources.

-Solar. Solar energy is an abundant source of energy that can be utilized in two ways: (i) to convert sunlight into electricity through a photovoltaic system and (ii) to generate heat using concentrating collectors. The estimated potential of the direct capture of solar energy is enormous. When solar energy strikes the Earth's atmosphere, approximately 30 % is reflected. After reflection by the atmosphere, Earth's surface receives about 3.9×10^{24} MJ incident solar energy per year, which is almost 10,000 times more than current global energy consumption. Thus, the harvesting of less than 1 % of photonic energy would serve all human energy needs [1]. Photovoltaic systems, as already discussed, are a novel approach to electricity generation as these use solar energy, which is freely available. Although the intermittent nature of solar radiation limits the use of this technology to some extent, for off-sunshine periods energy can be stored in a battery bank. Photovoltaic

Table 2.3 Different solar collectors with their operating temperatures, concentration factors, and power capacities

Solar collector	Concentration factor	Temperature (°C)	Power capacity
Flat-plate collector	1	<200	<1 MW (thermal)
Vacuum-tube collector	3	<300	<1 MW (thermal)
Concentrating solar collector (trough type)	40–80	<350	<50 MW (electrical)
Field mirror collector	200–700	<1,500	<150 MW (electrical)
Parabolic collector	1,000–2,500	<2,500	<100 kW (thermal)/E_{inh}

Modified from Brown et al. [20] and Friberg [21]

systems can be used not only as standalone systems but also connected to a grid to supply continuous electricity throughout the day. The efficiency of the solar cell typically ranges from 12 % to 15 % for a silicon solar cell. However, it is as high as 25–30 % for GaAs solar cells. The cost of the former is less as compared to the latter, and the latter is used mostly for space applications. The efficiency of the photovoltaic (PV) system can also be calculated from the product of the efficiencies of its various components such as the solar cell, module, and battery. From a health perspective, the potential benefits of solar energy applications seem very desirable. The two disadvantages of the PV technology can be low conversion efficiency and high cost of the solar cells, but these drawbacks can be overcome by intense research. On the other hand, solar thermal technology is at its maturity stage. Depending upon the temperature needed, different types of solar collectors can be used. Table 2.3 gives information about different solar collectors, their temperatures, concentration factors, and power capacities.

The flat-plate collector (FPC) is the simplest one: solar radiation incident on a flat transparent surface is transmitted to an equal-size absorbing/collecting surface generally composed of Cu or Al metal. Cu or Al metal is preferred because of high thermal conductivity (Cu) and comparatively reasonable cost (Al) of the material. Construction of a flat-plate collector is simple: various parts of a FPC are shown in Fig. 2.1.

A riser made of several metal tubes is attached to a black metallic surface called the receiver surface and placed between a metal box and a glazed surface. The metal box is thermally insulated by a suitable insulator (for example, glass wool). The glazed surface is exposed to sun to receive solar flux. To receive maximum solar flux, the metal box is tilted at an angle from the horizontal that is about the latitude of the location/city/village where it is being installed. The incident solar flux passes through the glazing gets absorbed on the receiver surface. The heat is transmitted to the water inside the riser, and the hot water goes up from the riser to a storage tank. The storage tank is connected to the riser from both ends, that is, top and bottom. Circulation of water inside the riser kicks in as soon as hot water rises up from the bottom to the top of the riser and goes to the storage tank by a combined effect of thermo-siphoning and gravity. It is important to note that the larger the area of receiver surface, the larger would be the thermal energy received. The concentration factor of a FPC is 1 and a thermal power up to 1 MW can be

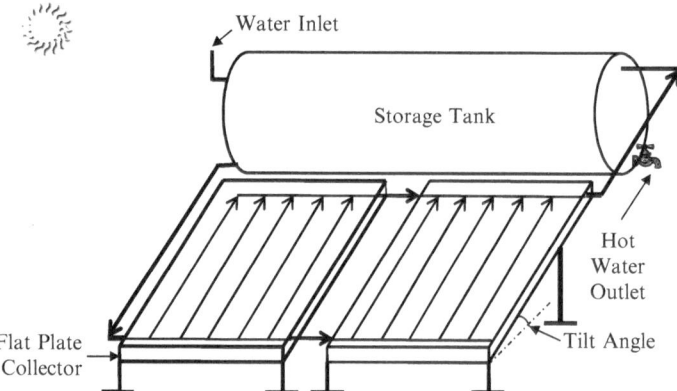

Fig. 2.1 A flat-plate collector system

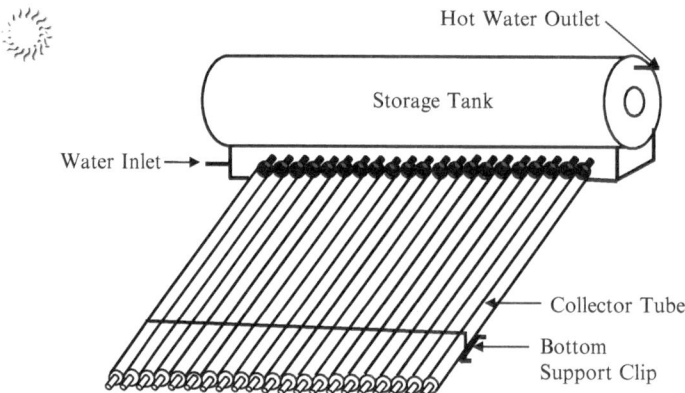

Fig. 2.2 Evacuated tube solar water heater

generated for a temperature range up to 200 °C by connecting flat-plate collectors in series. The other collectors shown are concentrating collectors. Their concentration factor, power, and operating temperature are higher. The application of solar energy in hydrogen production is discussed in the subsequent sections.

The vacuum-tube collector differs from the flat plate as it involves some tubes instead of a riser (Fig. 2.2). An evacuated tube is also shown in Fig. 2.3. An absorber plate gets heated when exposed to the sun transfers heat to a chemical via a heat pipe. The chemical tends to change phase from liquid to gas. Heat carried by the hot vapor/gas is then transmitted to water in the tank. Vacuum is created within the evacuated tube so as to minimize convective heat losses from absorber surface to ambient. The number of tubes can be increased or decreased depending upon the temperature of the hot water to be maintained. Some advantages include easy maintenance as tubes can be easily detached from the water heater.

Fig. 2.3 Evacuated tube

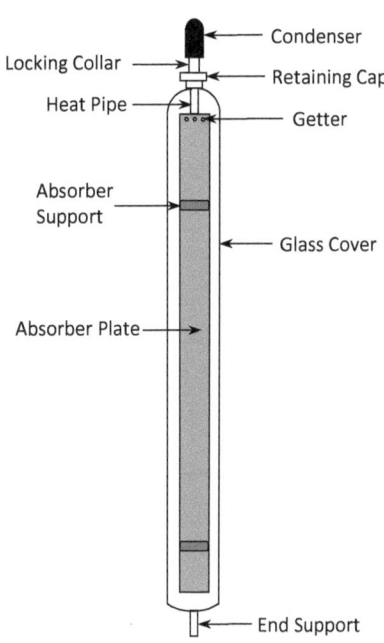

In the present book, the input energy source to produce hydrogen is taken to be solar energy; therefore, a brief introduction to solar energy is presented in the next section.

Table 2.4 Classification of geothermal sources

Temperature range	Application
Low (<90 °C)	Heating, cooling
Moderate (90–150 °C)	Heating, cooling, power generation
High (150–350 °C)	Heating, cooling, power generation, hydrogen production

Modified from Balta et al. [11]

-Geothermal. Geothermal energy is limited to appropriate geographic sites or locations where the resource is present; however, there are many such sites worldwide, spread over 24 countries with an operating potential of 57 TWh/year [22]. Geothermal energy is attractive for its ability to provide base load power 24 h per day. Extraction rates for power production will always be higher than refresh rates; reinjection helps restore the balance and significantly prolongs purpose only. Geothermal emissions are most significantly impacted by technology choices. Waste gases are more than 90 % CO_2 by weight [23], so if directly released, emissions will be high. Balta et al. [11] classified geothermal energy sources, based on temperature range for possible applications (Table 2.4).

It can be seen from Table 2.4 that the high geothermal resource temperature is about 350 °C, which is suitable for hydrogen production; however, recent research carried out by Landsvirkjun, Iceland's national power company, on deep drilling in

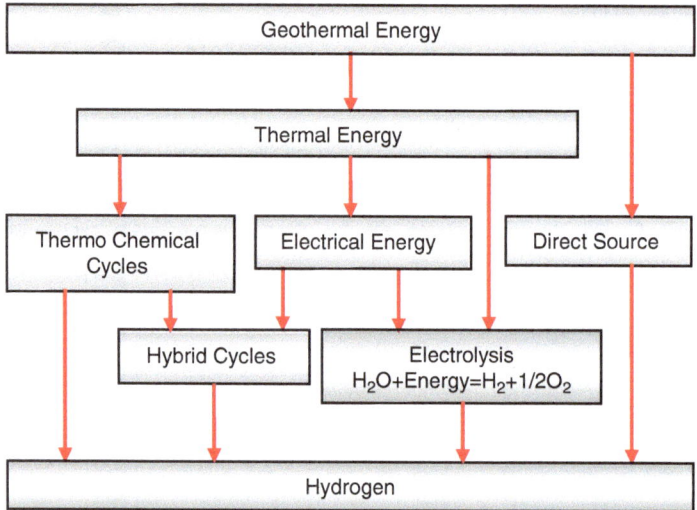

Fig. 2.4 Hydrogen production via geothermal energy (Modified from Balta et al. [11])

Iceland shows the possibility of extracting 500–600 °C of steam at a depth of 4–5 km for various applications, ranging from power production to hydrogen production. Presently, deep drilling is purely experimental, but it could become a possibility within the next decades [24–26]. Figure 2.4 shows various geothermal hydrogen production routes, which are mainly via thermal energy and direct application of geothermal energy.

Thermal energy application further depends upon the available temperature range. For example, if it is at a high temperature (350 °C), it can be used to provide heat in thermochemical cycles and hybrid cycles, and if it is in a moderate temperature range, electricity can be produced first, and it can then be used in electrolysis of water for hydrogen production. High temperature can also be used for high-temperature electrolysis of water whereas electricity can also be used in hybrid cycles. Some gases rich in hydrogen, for example, hydrogen sulfide (H_2S), also come from the geothermal well and can be used for hydrogen production. This route is shown as "direct source" in Fig. 2.4.

-Hydro. Hydroelectric power generation is an established technology that uses the potential energy of water to generate electricity. The main components of the hydropower plants are a dam/retaining wall, water turbine, and electrical generator. A dam or retaining wall is made across the width of a river so that the water level may rise on one side of the wall. On the other side of the dam/retaining wall, water turbines coupled with electricity generators are installed. The potential energy of water is then used to run turbines, and the turbines run generators and produce electricity. Water turbines are available in large variety, and selection depends upon the different water heads and flow rates. The Pelton wheel and Francis turbines are generally used for high water heads, and the Kaplan turbines can be used for low water heads.

Some intermediate water head turbines that can be used for both high and low water heads are Michel Banki and Deriaz turbines. The electricity produced is then supplied to the grid, from where it is distributed to its users. Mini hydro and hydel power stations can also be built to fulfill the electrical demands of a community living near small rivers and where the water head is not sufficient for a big hydropower plant. Hydro energy is essentially used to produce electricity, and then the electricity can be used for hydrogen production via electrolysis. Hydropower plants are more eco-friendly then thermal power plants as they cause less harm to the environment, but because these require very large civil structures and community relocation for those who live near the river, substantial public resistance sometimes occurs.

-**Biomass.** Biomass can also be used as an alternative as it has a large stored potential of renewable energy, which can be utilized to produce power by combustion or by thermochemical or biochemical conversion to liquid (ethanol, methanol) or gaseous fuels (methane, hydrogen) [27]. However, the inherent inefficiency of photosynthesis, which captures only a small percentage of solar energy reaching the Earth's surface, limits its usefulness as a major energy source [28]. Some high-yielding crops, for example, South American sugar cane, are already being used successfully as fuel sources, mainly for transport. Bioelectricity can be an important option in supporting electricity needs, particularly of rural populations in lower-income countries. The production of electricity using biomass has some health consequences, but these are much less than those from coal, oil, and natural gas. Wood sawdust and sugar cane bagasse are some general forms of biomass that can be used to produce electricity and hydrogen. Abudala et al. [29] analyzed exergetically a hydrogen production system based on biomass that uses wood sawdust and found the hydrogen yield reaches 80–130 g H_2/kg biomass. The biomass is introduced to a gasifier at an operating temperature range of 1,000–1,500 K. Also, a 4.5 kg/s steam at 500 K is used as the gasification medium.

-**Wind.** Wind mills and horizontal-axis and vertical-axis turbines are used to convert the kinetic energy of the wind into electricity. It is one of the more cost-effective forms of renewable energy with today's technology. The electricity produced by wind energy can be supplied to the grid. The technology is beneficial for locations where wind velocity is high, for example, the coastal and sub-coastal areas. For better functioning of a wind energy system, knowledge of the natural geographic variation in wind speed is important to smooth out fluctuations. Similar to the limitations of solar energy, wind energy generation is also affected by the intermittent nature of wind speed. Similar to hydro energy, wind energy also essentially used to produce electricity first and then the electricity can be used for hydrogen production.

-**Tidal, Wave, and Ocean Thermal.** Some other renewable sources are tidal, wave, and ocean thermal technologies that can produce electricity or can help reduce the electrical load of a power plant. Tidal energy utilizes the power of tide to produce electricity whereas wave energy systems use the waves formed in an ocean or sea. Oscillators are placed in the sea, and their oscillatory motion when waves come in contact with them is utilized to generate electricity. The ocean thermal technology uses the temperature difference between the upper and the deep lower layers of ocean water to generate electricity. The electricity produced by this technology may be utilized to produce hydrogen by electrolysis of seawater.

-Hybrid Renewable Systems. Hydrogen can also be produced by combining two or more renewable systems, for example, photovoltaic and wind [30, 31]. The two technologies are not competing with each other; rather, they are complementing and supporting each other. On one hand, the wind technology can be beneficial for off-sunshine periods; on the other hand, the solar photovoltaic technology can compensate for conditions of no wind during the daytime. This symbiotic behavior of the two technologies ensures a better and continuous supply of electricity to the electrolyzer to produce hydrogen. The excess power produced by the system can be stored in batteries and used in adverse conditions. Another example of the coupling of two technologies can be solar thermal and geothermal. The hot water from the geothermal sources can be further heated to a desirable temperature (approximately 550 °C) by using solar concentrating collectors; then, by using a high-temperature electrolyzer, hydrogen can be produced. One of the advantages of the hybrid renewable technology is to ensure a continuous supply of input energy, which when using these technologies individually sometimes can be challenging. The performance of such hybrid systems can be better than the systems that use the two technologies separately.

Chapter 3
Solar Energy Aspects

3.1 Solar Radiation

To utilize solar energy, the solar irradiance of the place where it is being used should be known, including the type of day and the number of sunshine hours or daylength. The days of a year can be divided into four types, namely, clear day (blue sky), hazy day, partially hazy and cloudy day, and completely cloudy day, based on the ratio of daily diffuse to daily global radiation and the number of sunshine hours (Singh and Tiwari [32]). Their criteria are given in Table 3.1.

The number of sunshine hours includes those hours for which the intensity of solar radiation was either equal to or more than 120 W/m^2. The usefulness of solar energy can be seen in terms of higher intensity of solar radiation, a lower ratio of daily diffuse to daily global radiation, and a long daylength or a large number of sunshine hours. Global radiation is composed of two components, beam and diffuse radiation. The types of days given here largely depend upon the following parameters: cloudiness/haziness factor, atmospheric transmittance, perturbation factor, and background diffuse radiation. The first two parameters influence beam radiation; the last two can be used to predict diffuse radiation. Some mathematical models that can predict the solar radiation are presented next.

Model 1. Model 1 predicts beam and diffuse radiation by using the foregoing parameters and some other parameters, namely, air mass and optical thickness of the atmosphere.

The terrestrial beam radiation (I_{HB}) received on a horizontal surface can be given as [32]:

$$I_{HB} = I_N \cos\theta_z = I_{ON} \exp[-(m. \, \varepsilon \, T_R + \alpha)]. \cos\theta_z \qquad (3.1)$$

I. Dincer and A.S. Joshi, *Solar Based Hydrogen Production Systems*,
SpringerBriefs in Energy, DOI 10.1007/978-1-4614-7431-9_3, © The Author(s) 2013

Table 3.1 Different types of days

Type of day	Ratio of daily diffuse to daily global radiation (R)	Number of sunshine hours (N)
a	R ≤ 0.25	≥ 9
b	0.25 ≤ R ≤ 0.5	7 ≤ N ≤ 9
c	0.5 ≤ R ≤ 0.75	5 ≤ N ≤ 7
d	0.75 ≤ R ≤ 1	N ≤ 5

Source: Singh and Tiwari [32]

where

I_N = normal terrestrial solar radiation (W/m^2)
I_{ON} = normal extraterrestrial solar radiation (W/m^2)
T_R = cloudiness/haziness factor
α = lumped atmospheric parameter for beam radiation
m = air mass (dimensionless)
ε = optical thickness of the atmosphere
θ_Z = zenith angle (degree)

Air mass can be defined as the ratio of the optical thickness of the atmosphere (ε) at a specific point in time to the optical thickness of atmosphere when the sun is at the zenith and can be given in terms of zenith angle as [33, 34]:

$$m = \left[\cos\theta_Z + 0.15 \times (93.885 - \theta_Z)^{-1.253}\right]^{-1}$$

and the optical thickness of atmosphere as

$$\varepsilon = 4.529 \times 10^{-4} \times m^2 - 9.66865 \times 10^{-3} \times m + 0.108014.$$

Rewriting Eq. 3.1 as

$$\frac{I_{HB}}{I_{ON} \cdot \cos\theta_z} = \exp[-(m.\varepsilon.T_R + \alpha)]$$

the foregoing equation after taking the log becomes

$$\ln\left[\frac{I_{HB}}{I_{ON} \cdot \cos\theta_z}\right] = T_R(m.\varepsilon) + (-\alpha) \tag{3.2}$$

Equation 3.2 is a standard straight-line equation as

$$Y = MX + C \tag{3.3}$$

where $Y = \ln\left[\frac{I_{HB}}{I_{ON}\cos\theta_z}\right]$; $X = m\varepsilon$; $M = -T_R$, and $C = -\alpha$

By using linear regression analysis of Y on X, regression coefficients M and C can be obtained and hence the cloudiness/haziness factor (T_R) and lumped atmospheric parameters for beam radiation (α) can be evaluated.

Again, the terrestrial diffuse radiation (I_{HD}) on a horizontal surface can be written as

$$I_{HD} = K_1(I_{ON} - I_N).\cos\theta_Z + K_2 \tag{3.4}$$

here, (I_{ON}) is given by

$$I_{ON} = I_{SC}[1.0 + 0.033 \times \cos(360 \times n/365)] \tag{3.5}$$

where K_1 = perturbation factor (dimensionless), K_2 = background diffuse radiation (W/m^2), I_{SC} = solar constant (1,367 W/m^2), and n = day of the year (for example, $n = 15$ for 15 January).

The cloudiness/haziness factor, lumped atmospheric parameter for beam radiation, perturbation factor, and background diffuse radiation for diffuse radiation for a particular month can be obtained by linear regression analysis for known hourly data of global and diffuse radiation and the average monthly value for the day of the year (n), given by Duffie and Beckman [35]. After knowing the cloudiness/haziness factor (T_R) and lumped atmospheric parameter for beam radiation (α) of Eq. 3.1 and the constants (K_1) and (K_2) of Eq. 3.4, the hourly beam and diffuse radiation on a horizontal surface can be calculated. Then, the total radiation for any inclination with any orientation of a solar thermal device can be evaluated by using the Liu and Jordan formula and can be given as [36]:

$$I_t = I_b R_b + I_d R_d + \rho.R_r(I_b + I_d) \tag{3.6}$$

where

$$R_b = \frac{\cos\theta_i}{\cos\theta_z}$$

$$R_d = \frac{(1 + \cos\beta)}{2}$$

$$R_r = \frac{(1 - \cos\beta)}{2}$$

$$\cos\theta_z = \cos\phi\ \cos\delta\ \cos\omega + \sin\delta\ \sin\phi$$

and

$$\cos \theta_i = (\cos\phi \cos \beta + \sin \phi \sin \beta \cos \gamma) \, \cos\delta \, \cos\omega + \cos \delta \sin \omega \sin \beta \sin \gamma$$
$$+ \sin \delta(\sin\phi \cos \beta - \cos \phi \sin \beta \cos \gamma)$$

where

ρ = reflectance
ϕ = latitude of the place (degree)
β = surface tilt angle from horizontal (degree)
γ = solar azimuth angle (degree)
δ = angle of declination (degree)
ω = hour angle (degree)
θ_i = angle of incidence of solar radiation (degree)
θ_z = zenith angle (degree)

Model 2. The terrestrial beam radiation received on a horizontal surface can also be determined as given next (by Perez [37]):

$$I_{HB} = I_{ON} \exp\left[\frac{-T_R}{0.9 + 9.4 \cos \theta_z}\right] \cos \theta_z = I_N \cos \theta_z \qquad (3.7)$$

If the atmospheric transmittance factor for beam radiation (α) for lumped atmosphere is introduced in Eq. 3.7, as suggested by Joshi and Tiwari [38], then Eq. 3.7 can be modified as

$$I_{HB} = I_N \cos \theta_z = I_{ON} \exp\left[\frac{-T_R}{0.9 + 9.4 \cos \theta_z} + \alpha\right] \cos \theta_z \qquad (3.8)$$

Here, Eqs. 3.1 and 3.8 are used to evaluate (T_R) with and without (α) for city and flat land conditions by using the simple regression analysis, and the results will be compared for beam radiation.

The hourly variation of terrestrial diffuse radiation on a horizontal surface can be rewritten as

$$I_{HD} = 0.33(I_{ON} - I_N). \cos \theta_Z \qquad (3.9)$$

The two constants, perturbation factor (K_1) and background diffuse radiation (W/m^2) (K_2) of Eq. 3.4 for diffuse radiation for both regions, are also evaluated by the same procedure.

For known hourly beam (Eqs. 3.1 and 3.8) and diffuse (Eqs. 3.4 and 3.9) radiation on a horizontal surface by the second model, the total radiation (global radiation) for any inclination with any orientation of solar thermal devices can be evaluated by using the Liu and Jordan formula [36].

All these models predict beam and diffuse radiation with reasonably good accuracy and hence can be applied for the prediction of solar radiation. Once beam and diffuse radiations are known, one can add the two results to obtain global radiation. The cloudiness/haziness factor, atmospheric transmittance, perturbation factor, and background radiation for some selected Indian cities, namely, Bangalore, Jodhpur, Mumbai, New Delhi, and Srinagar, are given in Appendix 1. Different types of days for those cities are also presented in Appendix 2. Meteorological data of solar radiation for the past 10–12 years have been analyzed to evaluate the cloudiness/haziness factor, atmospheric transmittance, perturbation factor, background radiation, and different types of days [39].

Chapter 4
Solar Hydrogen Production

4.1 Introduction

The common methods of hydrogen production impose many concerns regarding the decline in fossil fuel resources, CO_2 emission, and ecological impacts. Subsequently, all the downstream industries that consume hydrogen involve the aforementioned drawbacks and risks. Therefore, H_2 production technologies with almost zero greenhouse gas emissions are the ideal candidates to address the hydrogen supply issue. In one approach, biomass gasification can be utilized to release hydrogen and carbon monoxide. Considering the CO_2 adsorption characteristics during the photosynthesis process, this method is basically carbon neutral. Alternatively, water electrolysis using renewable power sources such as geothermal, wind, and solar cells can be utilized.

To be an economical and sustainable pathway, hydrogen should be produced from a renewable energy source, that is, solar energy. Photo-catalytic water splitting is the most promising technology for this purpose, because H_2 can be obtained directly from abundant and renewable water and sunlight by this process. If successfully developed with economic viability, this could be the ultimate technology to solve both energy and environmental problems altogether in the future [40–42].

Compared to hydrogen production methods based on fossil fuels, the high investment cost of solar hydrogen generation is a challenging issue. Production of hydrogen in a regenerative fashion such as photochemical splitting of water has been studied Yoon et al. [43]. Photo-catalytic water splitting has been introduced as an efficient and cost-effective way to produce hydrogen, in which sunlight is absorbed and water is split directly into hydrogen and oxygen. Many efforts in photo-catalytic water splitting have focused on increasing the efficiency and stability of the photoactive materials [44, 45] to achieve the required efficient target to be viable for commercialization [46].

The reductive side of this process requires the development of catalysts that promote the reduction of protons to molecular hydrogen, facilitated by direct excitation

I. Dincer and A.S. Joshi, *Solar Based Hydrogen Production Systems*, 27
SpringerBriefs in Energy, DOI 10.1007/978-1-4614-7431-9_4, © The Author(s) 2013

by a photo-sensitizer. Recently, molecular platinum- and palladium-based systems have been developed as heterogeneous catalysts [47, 48]. Supramolecular complexes in a photo-catalytic hydrogen production scheme that results in high turnover rates and numbers are promising [49].

Catalytic hydrogen evolution from a neutral aqueous solution using hydrogenase enzymes that possess iron or nickel cofactors shows turnover frequencies of 100–10,000 mol H_2 per mole of catalyst per second at their thermodynamic potential [50–52]. However, the large size of these enzymes and their instability under aerobic, ambient conditions has encouraged new searches for alternative molecular complexes to produce H_2 from water in a nonbiological setting. Despite the high catalytic activity of precious metals for water reduction at a low over-potential, high cost challenged their widespread application. Therefore, the main challenge in catalytic hydrogen production from water is to create Earth-abundant molecular systems with high catalytic activity and stability.

This aim is focused on efforts to develop water-splitting devices in which the energy inputs, which are either direct sunlight or solar-based electricity, handle a hydrogen-evolving reaction in an aqueous environment. In this book, photo-induced hydrogen production methods are introduced and explained. Solar-driven water splitting combines several attractive features for sustainable energy utilization. The conversion of solar energy to a type of storable energy has crucial importance. In the first part of this chapter, background information is presented regarding different routes of water dissociation with light energy to generate hydrogen. The photo-electrochemistry of water splitting is discussed, as well as photo-catalytic reaction mechanisms. The recent advances and new contributions to light-based hydrogen production systems are reviewed. The overall photo-catalytic water-splitting systems to produce hydrogen and oxygen are described from design and scale-up aspects. At the end, the principles of photo-electrolysis and photo-biological methods are also explained and detailed examples are presented.

4.2 General Classification of Solar Hydrogen Production

Hydrogen production using solar energy can be classified mainly into four types: (1) photovoltaic, (2) thermal energy, (3) photo-electrolysis, and (4) bio-photolysis. The thermal energy from solar energy can be utilized in two ways: low-temperature and high-temperature applications, also called concentrated solar energy. Photovoltaic, photo-electrolysis, and bio-photolysis are considered as low-temperature applications whereas solar thermolysis, solar thermochemical cycles, solar gasification, solar reforming, and solar cracking are high-temperature applications of concentrated solar thermal energy. Four major ways in which solar energy can be utilized to produce hydrogen are given in Fig. 4.1 [53]. Now, we discuss each hydrogen production method, one by one, in the following sections.

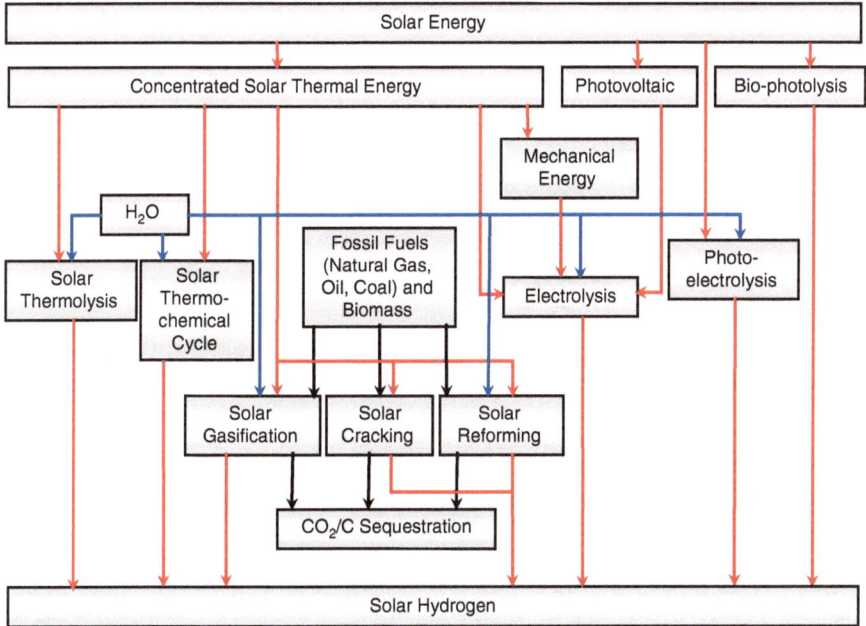

Fig. 4.1 Solar hydrogen production (Modified from Steinfeld [53])

4.2.1 Photovoltaic-Based Electrolysis

In the beginning of the 1970s, photovoltaic (PV) panels were utilized to produce hydrogen by electrolysis of water using electricity produced by the photovoltaic cells [10, 54–56]. The electrolysis of water can be carried out by a current generated by the photovoltaic cells. More extensive research is needed in this area, as the hydrogen produced by this technology is not cost effective because the PV technology is costly. The end product of the electrolysis was also oxygen. Figure 4.2 shows a schematic diagram of hydrogen production system that uses photovoltaic technology. The electrolysis of distilled water using electricity produced by the PV panel takes place in an electrolyzer unit and produces hydrogen and oxygen as the end product. One advantage with PV technology is that it does not emit greenhouse gases during the operation. From economic and ecological points of view, the most effective production method is considered the application of a photovoltaic current source (PCS) as a result of transforming radiant solar energy directly into electrical energy. When the efficiency of modern photoconverters (i.e., photovoltaic) and electrolyzers is about 20 % and 80 %, respectively [57], the total efficiency of solar radiant energy transformed to chemical hydrogen energy is nearly 16 % [10].

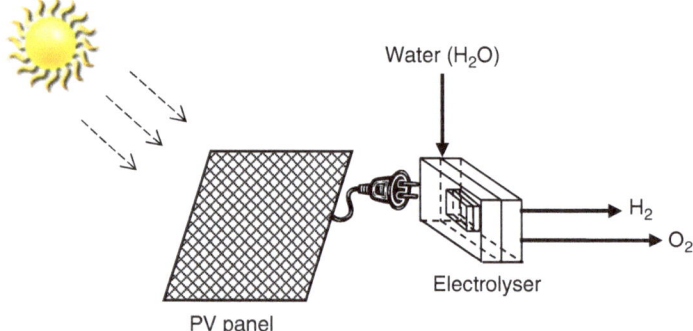

Fig. 4.2 Schematic diagram of photovoltaic hydrogen production system

4.2.2 Concentrated Thermal Energy-Based Hydrogen Production

Various thermochemical methods for solar hydrogen production are shown in Fig. 4.1. The chemical source of hydrogen, that is, water for solar thermolysis and solar thermochemical cycles, fossil fuels for solar cracking, and a combination of fossil fuels and H_2O for solar reforming and solar gasification are also shown in this figure. Because all these methods involve endothermic reactions, they make use of thermal energy of concentrated solar radiation as the energy source of high-temperature process heat (e.g., [53, 58]).

–Thermolysis. The single-step thermal dissociation of water, known as water thermolysis, can be given as

$$H_2O \rightarrow H_2 + 1/2O_2 \tag{4.1}$$

As already discussed, the reaction requires a high-temperature heat source, above 2,500 K, to have a reasonable degree of dissociation, and also requires an effective technique for separating H_2 and O_2 to avoid ending up with an explosive mixture. Among many effusion [59–61] and electrolytic separations [62, 63] are two ways to separate hydrogen from the products. Kogan [61] and Diver et al. [64] have tested semipermeable membranes based on ZrO_2 and other high-temperature materials at temperatures up to 2,500 K, but these ceramics usually fail to withstand the severe thermal shocks that often occur when working under high-flux solar irradiation. Some simple and workable methods are rapid quench by injecting a cold gas [65], by expansion in a nozzle, or by submerging an irradiated target in liquid water [66], but the quench causes a significant drop in the exergy efficiency and produces an explosive gas mixture. The very high temperature required by the process (e.g., 3,000 K for 64 % dissociation at 1 bar) poses severe material problems and can lead to significant re-radiation from the reactor, thereby lowering the absorption efficiency (e.g., [53]).

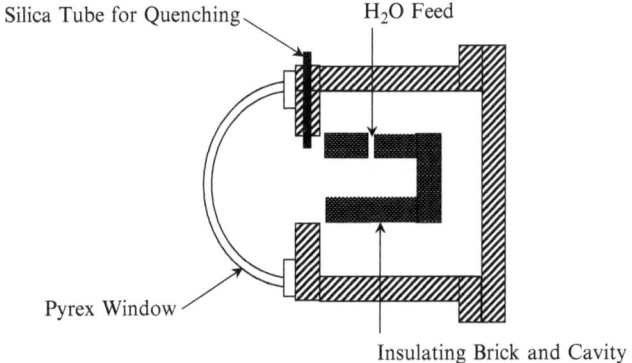

Fig. 4.3   An experimental setup for solar hydrogen production by thermolysis of water (Modified from Baykara [67])

–Solar Hydrogen Production Via Thermolysis of Water. As discussed, thermolysis involves a single-step decomposition of water molecules at high temperature (to 2,500 K). Because the process is reversible in nature, prevention of recombination of the products back to water is essential. The materials of water thermolysis reactor components are very special refractory materials that can withstand the very high temperature range of 1,500 K and are capable of enduring a chemically active environment. Also, the materials should withstand very high temperature gradients and swift temperature swings without degradation [67]. The various components of the solar reactor are shown in Fig. 4.3.

The cavity reactor is covered by insulating material (insulating bricks) on sides and back. A hemispherical glass window is attached to the front of the cavity. A silica tube is attached near the cavity, and a steam feed inlet is attached from the side of the cavity wall. When concentrated solar radiations focus on the cavity reactor, they are transmitted through the glass window: a hemispherical glass window is used to improve the concentration and minimize reflection losses. When the reactor temperature reaches 1,500 K or higher, steam is introduced to the cavity wall. Steam that comes in contact with the heated wall dissociates immediately into various products. As mentioned earlier, the thermolysis is a reversible process and recombination of products back to water should be prevented; therefore, an inert gas is passed through the silica tube for quenching. Quenching is a cooling process in which the cooling is done very quickly at a rate of 10^5–10^6 K/s. Quenching stops the reversible reaction by reducing the temperature abruptly and by diluting the concentration of the products. When the reacting gas mixture is quenched by means of cold auxiliary gas jets, it can retrieve up to 90 % of hydrogen within milliseconds. The overall efficiency of hydrogen production via high-temperature solar thermolysis of water requires the product of various efficiencies, namely, efficiency of the concentrator, efficiency of the receiver, efficiency of the solar reactor, and process efficiency of hydrogen production and the quenching process. Although the efficiencies of concentrator, receiver, and

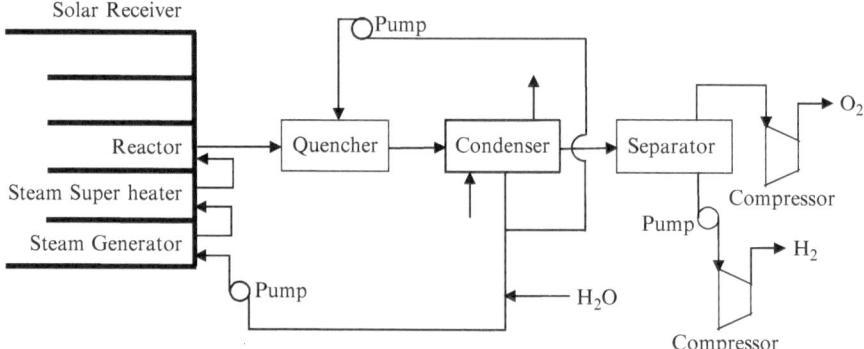

Fig. 4.4 Solar hydrogen production via thermolysis (Redrawn from Baykara [68])

reactor can be evaluated as was done in the previous case of hydrogen production via thermochemical cycles, the process efficiency of hydrogen production and quenching process efficiency depend upon the particular process chosen.

The overall chemical reaction can be given as

$$H_2O \rightarrow x_1H_2O + x_2OH + x_3O + x_4H + x_5O_2 + x_6H_2 \tag{4.2}$$

Here, the overall efficiency of the system depends upon the efficiencies of the system components and different processes (cycle efficiency). The product of all these efficiencies gives the overall efficiency. In the next chapter, calculations for concentrator, receiver, and reactor efficiencies are given. If thermolysis process, quenching, and separation process efficiencies are known, the overall efficiency of the system can be evaluated.

A simple flow diagram is shown in Fig. 4.4 to understand the hydrogen production via thermolysis [68]. As discussed earlier, the main components are solar receiver, reactor, quencher, condenser, separator, pumps, and compressors.

4.2.3 Thermochemical Cycles

Water-splitting thermochemical cycles have an upper edge on thermolysis because water splitting does not have the H_2/O_2 separation problem and further allows operating at relatively moderate upper temperatures (1,200 K) [53]. An efficient two-step thermochemical cycle using metal oxide redox reactions can be given as follows [69, 70]:

First step (solar):

$$M_xO_y \rightarrow xM + \frac{y}{2}O_2 \tag{4.3}$$

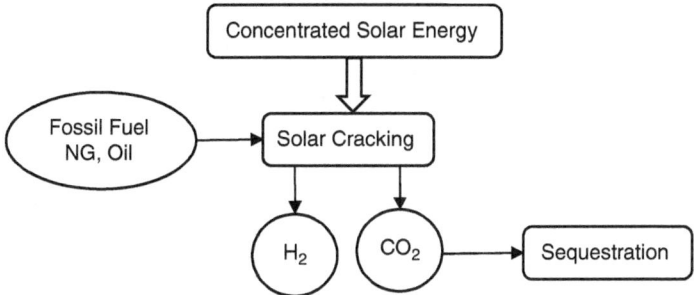

Fig. 4.5 Schematic diagram of solar cracking. *NG* natural gas (Modified from Steinfeld [53])

Second step (non-solar):

$$xM + yH_2O \rightarrow M_xO_y + yH_2 \tag{4.4}$$

where M denotes a metal and M_xO_y the corresponding metal oxide. The first, endothermic step is the solar thermal dissociation of the metal oxide to the metal or the lower-valence metal oxide. The second, non-solar, exothermic step is the hydrolysis of the metal to form H_2 and the corresponding metal oxide. The net reaction is ($H_2O = H_2 + 1/2O_2$), but because H_2 and O_2 are formed in different steps, the need for high-temperature gas separation is thereby eliminated [53].

4.2.4 Hydrogen Production by Decarbonization of Fossil Fuels

Hydrogen can be produced by fossil fuels by using solar thermochemical processes: solar cracking, solar reforming, and solar gasification. Figure 4.5 shows a schematic view of hydrogen production via solar cracking.

The process of solar cracking can be explained by the thermal decomposition of natural gas (NG), oil, and other hydrocarbons, and can be represented by the simplified net reaction:

$$C_xH_y \rightarrow xC(gr) + \frac{y}{2}H_2 \tag{4.5}$$

The steam reforming of natural gas (NG), oil, and other hydrocarbons, and the steam gasification of coal and other solid carbonaceous materials, can be expressed by the simplified net reaction:

$$C_xH_y + xH_2O \rightarrow \left(\frac{y}{2} + x\right)H_2 + xCO \tag{4.6}$$

Depending on the reaction kinetics and on the presence of impurities in the raw materials, other compounds may also be formed. Reaction 4.5 produces a carbon-rich condensed phase and a hydrogen-rich gas phase. The carbonaceous solid product can either be sequestered without CO_2 release or used as a material commodity or reducing agent under less severe CO_2 restraints. The product of reaction 4.6 is syngas. The CO content present in the syngas can be shifted to H_2 via the catalytic water–gas shift reaction ($CO + H_2O = H_2 + CO_2$), and the product CO_2 can be separated from H_2 using, for example, the pressure swing adsorption technique [53].

Some of these processes are already in industrial-scale use. The heat supplied by burning the feedstock as the process heat further results in either the contamination of the gaseous product, in the case of internal combustion, or in reduced thermal efficiency because of irreversibility associated with indirect heat transfer in the case of external combustion. Some advantages of using solar energy for process heat are that the discharge of pollutants can be avoided, the gaseous products are not contaminated, and the calorific value of the fuel is upgraded by adding solar energy in an amount equal to the enthalpy change or total energy required by the reaction. The steam reforming/gasification method requires additional steps for shifting CO and for separating CO_2 whereas thermal cracking accomplishes the removal and separation of carbon in a single step. In contrast, the major drawback of the thermal decomposition method is the energy loss associated with sequestration of carbon. Thus, solar cracking may be the preferred option for natural gas (NG) and other hydrocarbons with high H_2/C ratio. For coal and other solid carbonaceous materials, solar gasification via reaction 4.6 has the additional benefit of converting a solid fuel traditionally used to generate electricity by the Rankine cycle into a cleaner fluid fuel, that is, cleaner only when using solar process heat that can be used in highly efficient fuel cells [53]. Figure 4.6 shows a schematic view of hydrogen production via solar gasification/reforming.

4.2.5   Solar Cracking of Methane

Figure 4.7 shows a laboratory-scale prototype of a solar reactor that is used for solar cracking of methane. It comprises a graphite cavity receiver (blackbody absorber of cubic shape, 20-cm-side) and an aperture (9-cm-diameter) that allows concentrated solar radiation to enter into the cavity through a quartz window. The inside cavity of the reactor is swept by nitrogen, and is therefore separated from ambient oxidizing atmosphere. The cracking of methane takes place in the four tubular graphite zones set in parallel and vertically in the solar absorber. Each reaction zone has an independent gas inlet by which a mixture of Ar and CH_4 is fed to the reactor. The reactor tube is composed of two concentric graphite tubes. The inner tube is used for the gas inlet (12-mm outer diameter, 4-mm inner diameter) and the outer tube for the gas outlet (24-mm outer diameter, 18-mm inner diameter). The gas mixture of Ar and CH_4 enters the inner tube and flows out from the annular space between the outer and inner tubes. A cross-sectional view of the reactor is shown in

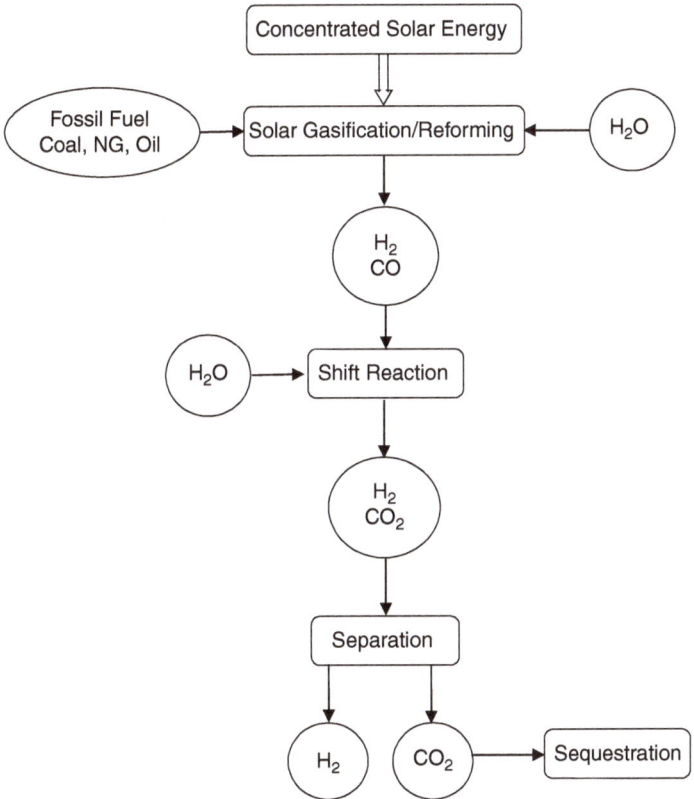

Fig. 4.6 Schematic diagram of solar gasification/reforming (Modified from Steinfeld [53])

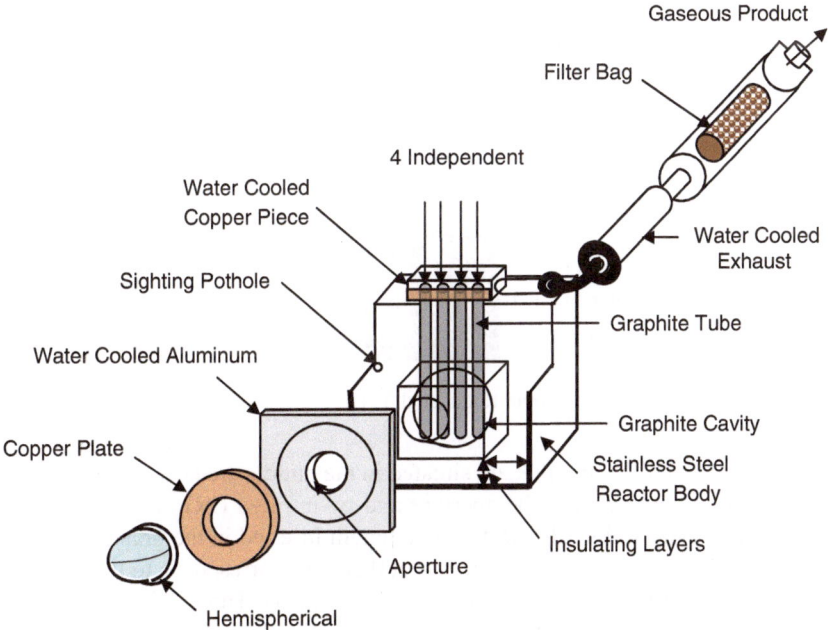

Fig. 4.7 Laboratory-scale setup of solar cracking of methane (Modified from Rodat et al. [71])

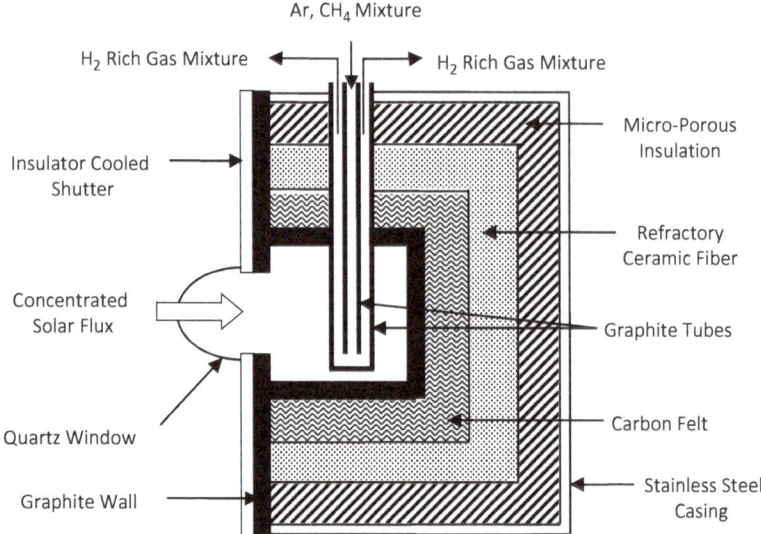

Fig. 4.8 Cross-sectional view of a 10-kW laboratory-scale solar reactor (Modified from Abanades and Flamant [72])

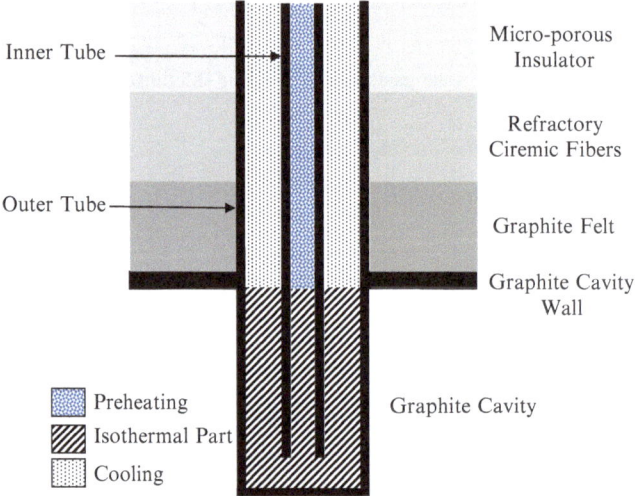

Fig. 4.9 Reactor tube (Modified from Rodat et al. [71])

Fig. 4.8. The graphite tubes that are situated in the middle are heated both by direct solar radiation coming from the aperture and by infrared (IR) radiation from the surrounding cavity walls. The heated tube length inserted in the graphite cavity is about 0.161 m; the remaining length (about 0.203 m) corresponds to the insulation zone (Fig. 4.9), based on the system in Rodat et al. [71]. Three different insulating layers surround the reactor cavity to lower conduction losses.

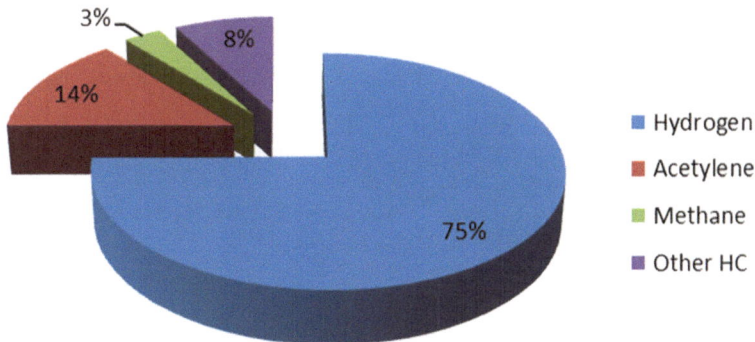

Fig. 4.10 Concentration of hydrogen in the gas mixture (Modified from Rodat et al. [71])

The insulating layer is formed by three different insulators, namely, graphite felt, refractory ceramic fiber, and microporous insulator. Each of the three insulating materials is 0.05 m thick, making the thickness of the insulating layers 0.15 m. It can be seen from Fig. 4.9 that the graphite felt is in contact with the cavity and the microporous insulator is the outermost insulator; between these two is the ceramic fiber. The thermal conductivity of the graphite felt is 0.46 $Wm^{-1} K^{-1}$, whereas it is 0.35 $Wm^{-1} K^{-1}$ (at 1,400 °C) for refractory ceramic fiber (62 % Al_2O_3, 30 % SiO_2), which is operated up to around 1,600 °C. For the microporous insulator (20 % ZrO_2, 77.5 % SiO_2, 2.5 % CaO), which has an operating temperature of about 1,000 °C, the thermal conductivity is 0.044 $Wm^{-1} K^{-1}$ (at 800 °C). The surrounding outer shell/case of the reactor is made of stainless steel (535 mm × 535 mm × 373 mm) and the front face is made of aluminum with water-cooling channels. The hemispherical quartz window is supported by a water-cooled copper plate. The transparent window closes the reactor cavity to maintain an inert atmosphere, and it also prevents the graphite from oxidizing as the graphite cavity is swept by a nitrogen flow. The reactor is set up at the focus of the 1-MW solar furnace of the CNRS-PROMES laboratory and is designed for a nominal power of 10 kW [71]. This solar furnace has a field of 63 heliostats for full power (45 m^2 per heliostat) and a parabolic concentrator (1,830 m^2) that can deliver up to 9,000 suns at the focal plane. During experiments at the 10-kW scale, only a fraction of the parabola is used by limiting the number of heliostats tracking the sun and by using a shutter [71].

Rodat et al. [71] have conducted experiments to produce hydrogen from solar cracking of methane (Fig. 4.10). The first experimental step was heating the reactor under an argon flow in the tubes. Once the desired temperature was reached, the mixture of argon and methane was injected with a controlled composition.

Cracking of methane takes place at the graphite tube reactor, and then the decomposed gas mixture passes through a bag filter where the carbon particles become settled. When flowing in the tube, gases pass through three different zones consisting of a pre-heating zone, an isothermal zone, and a cooling zone (see Fig. 4.9). With the help of a gas-liquid chromatograph (GLC), the gas mixture

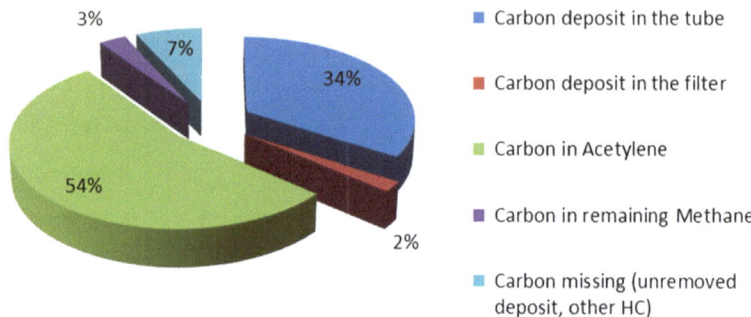

Fig. 4.11 Concentration of carbon. *HC* hydrocarbons (Modified from Rodat et al. [71])

(CH_4, C_2H_6, C_2H_4, C_2H_2, and H_2) was analyzed for the concentration of each component (Figs. 4.10, 4.11). The chromatographic analysis was based on thermal conductivity detection, and the carrier gas was argon, also used as buffer gas during methane cracking experiments.

Rodat et al. [71] conducted a series of five experimental runs at 1,923 K with a constant argon flow rate of 18 NL/min (3×10^{-4} Nm^3 s^{-1}) and a methane flow rate varying from 2 to 10 NL/min (3.33×10^{-5} to 1.67×10^{-4} Nm^3 s^{-1}) and established a mass balance of different gases (Fig. 4.11). It is clear from Fig. 4.10 that the concentration of hydrogen in the gas mixture is 75 %, acetylene is 14 %, methane is 3 %, and the remaining 8 % is thus attributed to other hydrocarbons. According to the foregoing mass balance, methane is efficiently converted into H_2 (75 % H_2 yield). The concentration of carbon, as shown in Fig 4.11, is highest in acetylene with a total of 54 %, and only 36 % of the carbon is recovered as solid carbon either in the reactor tubes (34 %) or in the filter (2 %). Three percent of carbon stays in the unconverted methane. The remaining part (7 %) may be contained in other hydrocarbons or may correspond to unremoved solid carbon. From Figs. 4.10 and 4.11 it can be said that methane is not as well converted into solid carbon as it is in H_2 because of the substantial presence of acetylene.

Another laboratory-scale prototype for hydrogen production via methane cracking is the experimental system of Abanades and Flamant [72] (Fig. 4.12).

This system is composed of a vertical-axis solar furnace, a solar reactor, a gas–solid separation step (filtering device), and a gas analysis system for measuring species (mixture gas constituents) concentration. The design parameter involves a tubular graphite receiver (10-mm inner diameter, 17-mm outer diameter, and 61-mm-long nozzle) set vertically that absorbs concentrated solar radiation through a glass window. Carbon felt 30 mm thick is used as the insulation layer surrounding the inner graphite tube. A mixture of argon and methane as the reactive gas is injected and dissociates in the graphite nozzle directly exposed to the high flux solar irradiation. Thus, the gas is heated mainly by conductive, convective, and radiative heat transfer from the hot wall. Carbon particles are formed and act as radiant heat absorbers; as they are exposed to concentrated solar flux, they also act as a

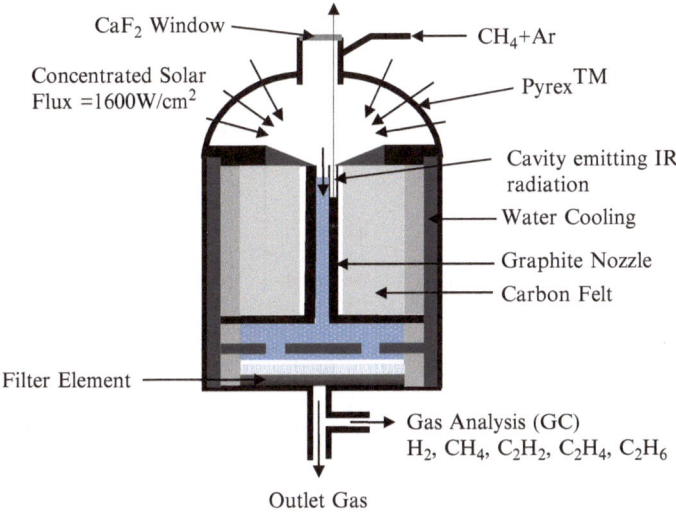

Fig. 4.12 A small-scale chemical reactor for dissociation of methane. *IR* infrared (Modified from Abanades and Flamant [72])

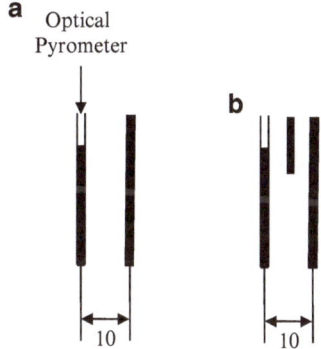

Fig. 4.13 Geometry of different graphite nozzles (Modified from Abanades and Flamant [12])

nucleation site for the reaction. Argon, the carrier gas, is injected at the top of the window to make the reactor atmosphere inert. It also helps to transport carbon particles formed by the reaction out of the reactor, hence preventing both particle deposition on the window and particle clogging. A filtering device is used at the reactor outlet to separate the carbon nanoparticles from the gas flow. From the reactor outlet, the gas mixture is sent to an on-line gas analysis system (gas chromatography, GC), which measures the concentration of different components, namely, H_2, CH_4, C_2H_6, C_2H_4, and C_2H_2. Abanades and Flamant [72] also found that the maximal chemical conversion in this small-scale solar tubular reactor was higher than 75 % [72]. Further, Abanades and Flamant [12] conducted experiments on two different types of graphite nozzles (Fig. 4.13) and concluded that the increase in wall

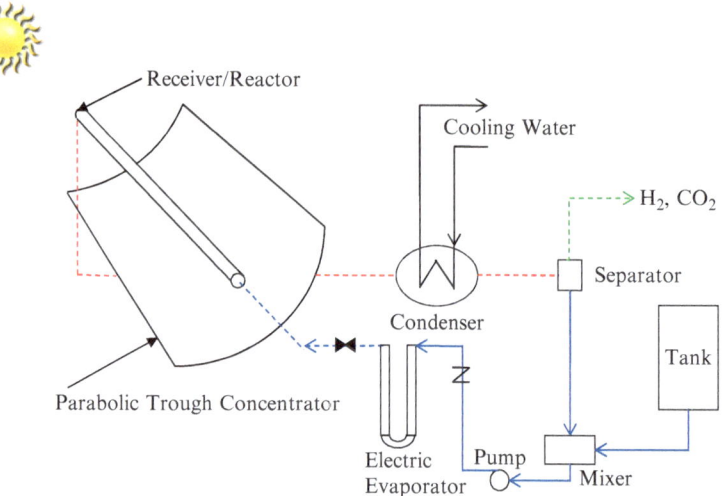

Fig. 4.14 Solar hydrogen production system with single tracking parabolic trough concentrator (Modified from Liu et al. [13])

temperature causes an increase in the conversion rates. In their experiment the maximum measured temperature was 1,540 °C; the methane conversion was about 75 % and the corresponding hydrogen yield was 65 % for the type "a" nozzle. For type "b," the methane conversion and the hydrogen yield were about 95 % and 82 %, respectively, when methane was injected at 0.1 L_n/min, and it decreased with time because of the temperature drop caused by the endothermic reaction.

4.2.6 Steam Reforming of Natural Gas

As an example of solar reforming, Fig. 4.14 shows a flow diagram of the experimental setup of Liu et al. [13]. A moderate temperature (up to 300 °C) thermochemical cycle using methanol was adopted for the production of hydrogen. The experimental setup consisted of an electric evaporator, a solar receiver/reactor, a water-cooling system, and a solar parabolic trough concentrator. Experiments were conducted on a 5-kW solar receiver/reactor composed of a linear one-track parabolic trough concentrator and a packed tubular bed receiver/reactor.

The single tracking parabolic trough concentrators (4 m long and 2.5 m wide) composed of silver-plated glass mirrors were positioned in an east–west direction. Further, the aperture area was 10 m^2 and the concentration ratio was 70; the concentrating collectors were capable of delivering up to 5 kW with a peak solar flux of 1,000 W/m^2. The fresh methanol liquid fuel and the feeding water were initially mixed in a molar ratio of 2.5:1 and fed to the electric evaporator with the aid of the pump. An electric evaporator was used to generate the superheated

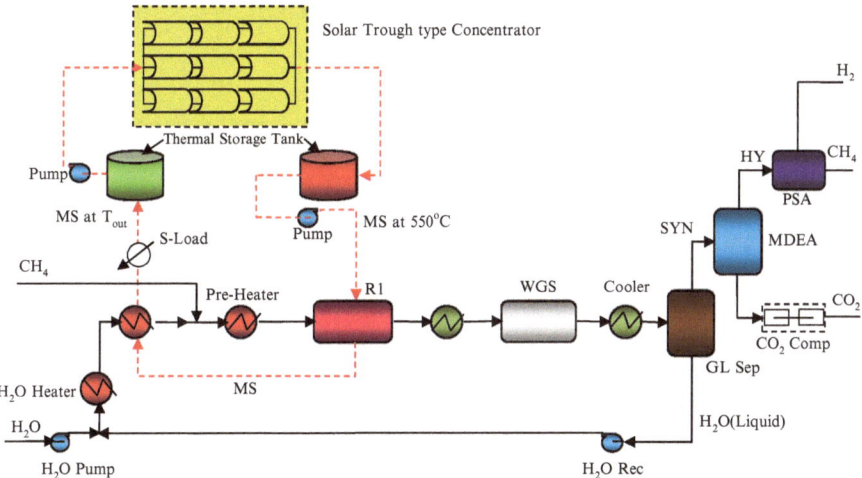

Fig. 4.15 Steam methane reforming using molten salts as thermal carrier and storage. *GL* gas–liquid (Modified from Giaconia et al. [73])

vapors of methanol and water. The vapors were allowed to flow into the mid-temperature solar receiver/reactor through a pipe that was heated by electric elements to keep the flow isothermal. Methanol steam reforming was carried out in the solar receiver/reactor using solar thermal energy at 150–300 °C. The gas mixture at the outlet of the solar receiver/reactor, which mainly included H_2, CO_2, non-reacted CH_3OH, and H_2O, was allowed to cool down first in a water-cooled plate fin condenser; then, via a separator, H_2 and CO_2 were separated from the nonreacted methanol and water. These nonreacted components were again sent to a mixer and then pumped to the electric evaporator. Liu et al. [13] have concluded that the chemical conversion of methanol can reach levels higher than 90 %, and the volumetric concentration of hydrogen in the gas products can account for 66–74 % above the solar flux of 580 W/m^2. The obtained maximum hydrogen yield per mole of methanol is 2.65–2.90 mol, approaching the theoretical maximum value, and the experimentally obtained thermochemical efficiency of solar thermal energy converted into chemical energy is in the range of 30–50 %, which is competitive with other high-temperature solar thermochemical processes, they concluded [13].

Another example of solar reforming is shown in Fig. 4.15. Giaconia et al. [73] have investigated methane steam reforming using molten salts that can be used as heat carrier and storage and suggested that the hydrogen selective membranes can be used to drive the reforming reaction at lower temperatures than conventional (less than 550 °C), with hydrogen purification achieved thereby.

At present, steam methane reforming (SMR) is an often-used, low-cost method to produce hydrogen. SMR is represented by following two catalytic chemical reactions in the gas phase:

$$CH_4 + H_2O \rightarrow CO + 3H_2 \tag{4.7}$$

$$CO + H_2O \rightarrow CO_2 + H_2 \tag{4.8}$$

Adding these two reactions (Eqs. 4.7 and 4.8), the overall reaction becomes

$$CH_4 + 2H_2O \rightarrow CO_2 + 4H_2 \tag{4.9}$$

The steam reforming reaction as represented by the first reaction is an endothermic reaction and requires a high temperature (800–1,000 °C). The second reaction is exothermic in nature, called a shift reaction, and requires a comparatively low temperature (200–450 °C). The water–gas shift reaction is mainly used to maximize the hydrogen yield unless the synthesis gas is the desired end product (e.g., in methanol production) [73]. The thermal energy for the reforming and shift reactions is generated by a concentrating solar power (CSP) plant, which consists of a solar collector field, a receiver, a heat transfer fluid loop, and a heat storage system.

The solar energy is concentrated through trough-type concentrators. The system has a heat storage arrangement using molten salt, a mixture of $NaNO_3$ and KNO_3 60/40 (w/w) in concentration, also known as "solar salt." The molten salt acts as the heat transfer fluid and removes the high-temperature solar heat from the receiver. The removed heat is then stored in an insulated thermal storage tank that maintains a temperature of 550 °C. From the thermal storage tanks, heat is supplied via molten salts by pumps to the heat users such as steam generators and endothermic reactors, where it releases its sensible heat. Finally, the heat carrier fluid is stored into a lower-temperature tank to restart the solar heat collection loop during sunshine (insolated) hours. The receiver pipe is made of two concentric cylinders. The outer cylinder, of transparent glass, is separated by a vacuum gap from the inner steel cylinder, which carries the heat transfer fluid (molten salt). The external glass cylinder acts a protective casing whereas the internal steel cylinder absorbs solar energy, which is converted into high-temperature sensible heat of the molten salts. The steel pipe is covered by a spectrally selective coating compound to maximize the efficiency of the receiver tube at higher temperatures (550 °C) and also to ensure the maximum absorption of solar radiation and minimize heat loss caused by infrared emissions from hot tubes [73]. The receiver pipe is located at the focal point of the concentrator and designed to withstand a high temperature of 580 °C.

Figure 4.15 shows the solar steam methane reforming (SMR) plant flow sheet. The plant comprises a compact tubular heat exchanger reactor (R1), which is a counterflow shell and tube-type heat exchanger. On the shell side the molten salt is flowing, and the inner tube packed with the catalyst is fed by the CH_4/H_2O gas mixture heated to about 500 °C (in pre-heater, PRE-HTR). As also stated previously, the molten salts from the hot storage tank of the solar plant have an inlet temperature of about 550 °C and cool down to lower temperatures (e.g., 530 °C) by transferring the sensible heat to the reacting mixture flowing in the inner tubes. The reaction temperature is maintained at 550 °C. The product mixture composed of $CH_4/H_2O/H_2/CO/CO_2$ is finally cooled in a cooler unit and purified. The water–gas shift reactor (WGS) is used to convert CO to CO_2 at a comparatively lower operating temperature (less than 450 °C), and the residual water is then condensed and separated in a gas–liquid separator (GL-SEP). The gas phase is treated in a CO_2 capture unit (e.g., by adsorption in amine solutions, MDEA) where CO_2 is separated and, finally, pure hydrogen can be optionally separated from nonreacted methane by a pressure swing absorber (PSA) unit. The removed CO_2 can be first liquefied (at pressures greater than 100 bar) and then sent to suitable stable disposal sites. The heat from the molten salts exiting the reforming reactor (R1) at about 530 °C can further be utilized to generate steam in the steam generator (STEAM-GN) unit and also to power other units such as the CO_2 removal and PSA units (S-LOAD). The resulting exit temperature of molten salt stream from the chemical plant is usually in the 450 °–510 °C range, and it then goes to the so-called cold tank of the CSP plant. It should be noted that the minimum temperature of molten salts for practical management is about 290 °C. Further, the sensible heat of the downstream molten salts can be used for electric power generation, heating, or water desalination, etc. If methane conversion on a single pass through the SMR reactor (R1) is not satisfactory or sufficient, then different reaction stages in series can be applied (R1, R2,...) for higher methane to hydrogen conversion, and, by deploying more membranes (M1, M2,...) set between catalyst beds, more hydrogen removal can be achieved. By so doing, a thermodynamic driving force resulting from hydrogen product removal enhances methane conversion to higher values despite the low temperatures [73].

4.2.7   Solar Gasification of Fossil Fuels

A solar gasification reactor is shown in Fig 4.16. The main components are a cylindrical cavity-receiver that contains a circular opening, the aperture, to let in concentrated solar radiation through a transparent quartz window. The cavity-type geometry is designed to capture the incident solar radiation effectively. The apparent absorptance approaches that of a blackbody absorber. A slurry of coke particles and water is injected at the front of the cavity through tangential injection nozzles and forms a vortex-type flow that progresses toward the rear side of the cavity, following a helical path, shown by dotted lines in Fig. 4.16.

A transparent glass window is attached to the reactor to minimize heat losses caused by radiation. The coke particles are directly exposed to the high flux solar irradiation, providing efficient heat transfer directly to the reaction site where the energy is needed for gasification, thereby bypassing the limitation imposed by conductive heat transport through the reactor walls. Further, the energy absorbed by the reactants is used to generate and superheat the steam and also to raise the reactant temperature to above 1,300 K and drive the gasification reaction [74].

Another type of solar gasification reactor (Fig 4.17) is a packed-bed solar reactor for coal gasification. In contrast to the other reactors, it is placed at the ground surface, and the concentrated solar radiations via a heliostat field mirrors are

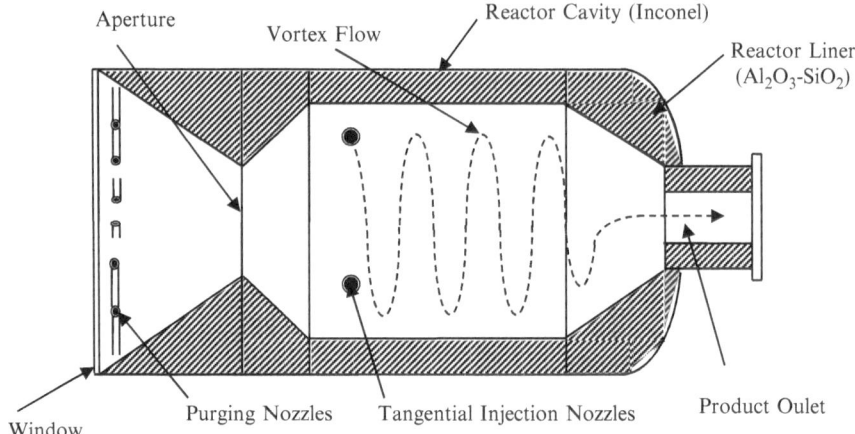

Fig. 4.16 A simplified diagram of a solar gasification reactor (Modified from Z'Graggena and Steinfeld [74])

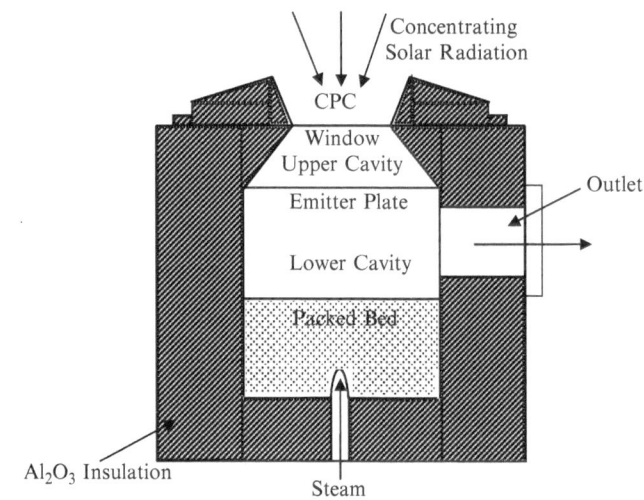

Fig. 4.17 Solar gasification reactor (Modified from Piatkowski and Steinfeld [75])

directed to the reactor [75]. A vertical-type reactor receives concentrated solar energy that is reflected by a hyperbolic reflector situated at the top of a solar tower which receives concentrated solar radiations from a heliostat field. There are two cavities in the reactor. The upper one functions as the solar absorber and contains a small windowed opening, the aperture, to let in concentrated solar radiation, whereas the lower one functions as the reaction chamber and contains the packed bed of coal on top of the steam injector.

The two cavities are separated by an emitter plate. A three-dimensional (3D) compound parabolic concentrator (CPC) is applied at the aperture of the reactor to further augment the incident solar flux before passing it through a quartz window into the upper cavity. Thus, in this arrangement, the emitter plate is directly irradiated and acts as a radiant absorber and radiant emitter to the lower cavity. The main purpose of the emitter plate is to eliminate direct contact between the quartz window and the reactants/products, preventing deposition of particles or condensable gases and assuring a clean window during operation. It further provides uniform heating of the bed through re-radiation. There is a possibility of inducing thermal shocks in the reactor because of intermittency of concentrated solar radiation, but the upper cavity also serves as a thermal shock absorber. Further, the reactor is operated in batch mode because the packed bed shrinks as the gasification reaction advances [75].

4.2.8 Hydrogen Production Through Carbonization of Bituminous Coal

Another method for hydrogen production may be the high-temperature (approximately 2,000 °C) carbonization of a suitably blended mixture of caking and non-caking varieties of bituminous coal. Figure 4.18 shows a schematic of hydrogen production via coke and coal gas, also known as coke oven gas (COG). Coke is formed by heating coal, also known as carbonization of coal at high temperature in absence of air. When steam is allowed to pass over the red-hot coke it produces a mixture of carbon monoxide and hydrogen, also known as water gas when both are in equimolar proportion. This water gas is further treated with excess of steam, which converts carbon monoxide to carbon dioxide, and after separation and carbon dioxide sequestration hydrogen is obtained. This process is shown on the left side of Fig. 4.18. Other than coke, vapors of volatile matter are produced. With some suitable methods the removable compounds of the volatile matter are eliminated and an uncondensed gaseous mixture, that is, coal gas, remains. The coal tar is removed by cooling and condensation. Ammonia is eliminated by scrubbing the residual gaseous mixture with water or dilute sulfuric acid solution. Then, naphthalene is removed by intensive cooling and benzol fractions are recovered by scrubbing with cresote oil. Hydrogen sulfide is removed by passing over ferric oxide. The residual uncondensed gaseous mixture is called coal gas, the compositional data of which are given in Table 4.1. Coal gas is rich in hydrogen, and it can be further enriched by carrying out carbonization of coal at temperatures above 2,000 °C. This method may be an alternative route for hydrogen production other than the conventional water–gas shift reaction route shown in Fig. 4.18.

Comparing hydrogen production with steam gasification of coal via synthetic gas or water gas and coal gas by the same amount of coal, the hydrogen yield would be higher with the coal gas. The molar percentage of hydrogen is competitive in

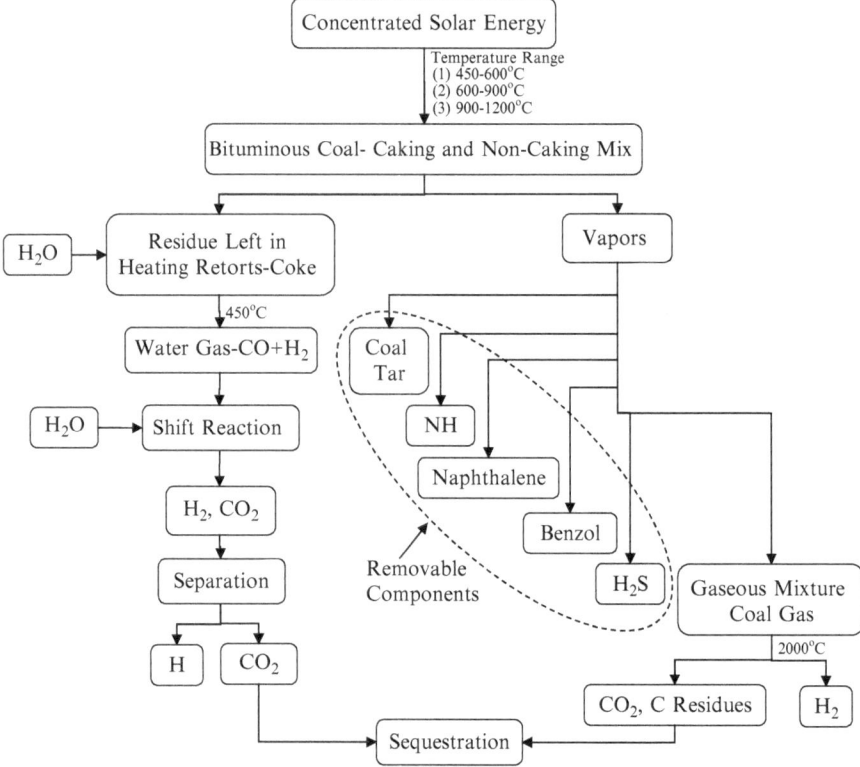

Fig. 4.18 Alternative scheme of hydrogen production through carbonization of suitably blended mix of caking and non-caking bituminous coal

Table 4.1 Composition of water gas, coal gas, and coke oven gas (COG)

Composition	Water gas (%) [76]	Coal gas (%) [76]	COG (in molar %) [77]
CO	40–42 %	7 %	8.7 %
H_2	48–51 %	47 %	57.3 %
CO_2	3–5 %	1 %	2.8 %
CH_4	0.1–0.5 %	32 %	20.5 %
N_2	3–6 %	4 %	6.7 %
C_2H_2	–	2 %	–
C_2H_4	–	3 %	–
C_2H_6	–	–	2.4 %
O_2	–	–	1.7 %
Other	–	4 %	–

both water gas and coal gas/COG whereas carbon monoxide is higher in the former. The reaction temperature of the former is low (458 °C) as compared to the latter (900–1,200 °C), and in both reactions the carbon dioxide or other carbon residues would be in the end product, which needs to be separated and sequestered.

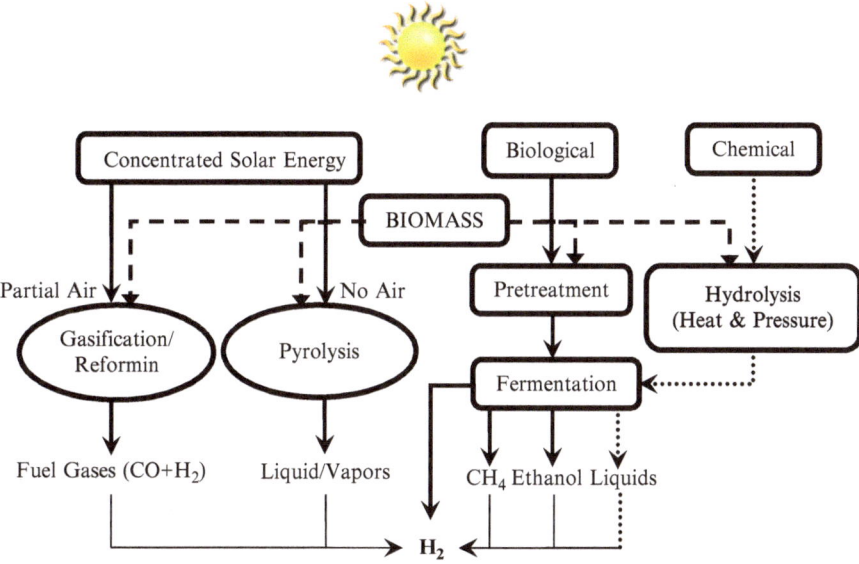

Fig. 4.19 Solar hydrogen production via biomass (Modified from Turner et al. [78])

Variation in the yield of coal gas/COG as reported is probably dependent on the quality of coal and the carbonization/oven temperature. If the temperature of the coal gas/COG is further increased to about 2,000 °C, the remaining hydrocarbons (i.e., C_2H_2, C_2H_4, CH_4, C_2H_6, etc., that comprise 25–37 M %) decompose, thereby further increasing hydrogen in the gaseous mixture. This high-temperature decomposition is not suitable for water gas as the remaining hydrocarbon (i.e., CH_4) is very much less (less than 0.5 %). The composition of water gas is also given in Table 4.1 for comparison [76, 77].

Hydrogen production via high-temperature decomposition of coal gas is not in commercial use, but this process may produce hydrogen at a comparable or higher rate than conventional steam gasification of coal using synthetic gas/water gas. By a proper combination of solar concentrator and solar reactor, this technology can be implemented for hydrogen production. Because of the higher reaction temperature, however, the problem associated with the materials cannot be avoided; therefore, more intensive research is required to make this technology commercially viable.

4.2.9   Solar Hydrogen Production Via Biomass

As discussed earlier, biomass can be a good alternative to coal as it comes under renewables. Different routes for hydrogen production via biomass [78] are shown in Fig. 4.19. The high-temperature thermochemical processes, namely, gasification,

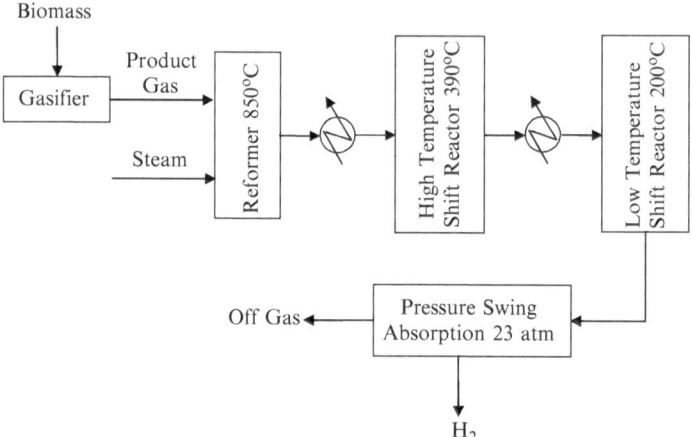

Fig. 4.20 Hydrogen production via biomass gasification (Modified from Koroneos et al. [79])

reforming, and pyrolysis, use concentrated solar energy whereas low-temperature processes such as biological and chemical hydrogen conversion use solar energy only.

The steam gasification/reforming of biomass can produce hydrogen-rich gas. The process is similar to the coal gasification shown in Fig. 4.6. The schematic diagram of hydrogen production via gasification and steam reforming of biomass (Fig. 4.20) describes the various components of hydrogen production via gasification of biomass. First, the biomass is gasified in a gasifier that produces product gas, then the product gas goes to a reformer unit for steam reforming at 850 °C. Then it passes through two shift reactors, namely, high- and low-temperature shift reactors, where the shift reaction takes place and through the pressure swing absorption technique, hydrogen is separated from the off-gas. The thermal needs of these units can very well be fulfilled by concentrated solar energy. However, the off-gas can also be used to supply heat to the steam reforming process [79]. The composition of hydrogen-rich product gas and the operation parameters are presented in Table 4.2 [80], which clearly shows the hydrogen percentage is 56.3, which is almost equal to the hydrogen percentage in the coke oven gas (57.3 %) and higher than water gas (47–51 %) (see Table 4.1).

The gasification of biomass in presence of oxygen or air, followed by water–gas shift, is a simple way to produce hydrogen. On the basis of the following reactions and typical composition for biomass, the stoichiometric yield of hydrogen produced by partial oxidation is 14.3 wt% [78]. The theoretical yield of hydrogen via steam gasification of lignocellulosic biomass, however, is 17 wt% [78]. The difference in the practical and theoretical yields is because in reality a small percentage of biomass carbon is converted in the first phase to char, tar, and CO_2, resulting in less CO available for the production of hydrogen by the second-phase water–gas

Table 4.2 Operation parameters and product gas composition

Operation parameters		Product gas composition (vol %, dry basis)	
Biomass feed rate (kg/h)	266.7	H_2	56.3
O_2 flow (m^3/h)	68.7	CO	8.9
Steam rate (kg/h)	45.8	CH_4	2.3
Temperature in the gasifier (°C)	800	CO_2	28.1
Catalyst	Z409R	N_2	4.2
Temperature in the CO-shift reactor (°C)	600	C_2 (C_2H_2, C_2H_4, C_2H_6)	0.2
Catalyst lifetime (h)	250	Gas flow (Nm^3/h)	427
WHSV (1/h)	2.5	LHV (MJ/Nm^3)	8.24
Cold gas efficiency (%, based on LHV)	69.9	–	–
Energy conversion efficiency (%, calculated by H_2 yield)	51.5	–	–

Source: Lv et al. [80]
LHV stands for "Lower Heating Value"

shift [78]. The chemical reaction of hydrogen production through gasification is expressed as follows:

$$CH_{1.46}O_{0.67} + 0.16O_2 \rightarrow CO + 0.73H_2$$
$$\text{Biomass} \qquad\qquad \text{Syngas} \qquad\qquad (4.10)$$

$$CO + H_2O \rightarrow CO_2 + H_2 \qquad\qquad (4.11)$$

The overall reaction becomes

$$CH_{1.46}O_{0.67} + 0.16O_2 + H_2O \rightarrow CO_2 + 1.73H_2 \qquad\qquad (4.12)$$

A two-step process, namely, pyrolysis of biomass followed by catalytic steam reforming of pyrolysis vapors or liquids (bio-oil), is another route to produce hydrogen from biomass [78]. Pyrolysis, in simple words, is a thermal decomposition process that occurs in an inert atmosphere. The maximum production of volatile intermediate compounds depends upon the high heat transfer systems. Such approaches are known as 'fast pyrolysis.' High yields (70–75 wt%, including water) of bio-oil can be achieved at 500–600 °C and residence times of approximately 1 s, with high heat transfer rates. Bio-oil is a mixture of carboxylic acids (mainly acetic and formic), aldehydes, alcohols, and lignin-derived methoxyphenolics (denoted as pyrolytic lignin), present as a low to medium molecular weight material that precipitates by adding water. Comparing pyrolysis with the traditional gasification reveals several advantages of the former at small-scale arrangements.

- Bio-oil has a higher volumetric energy density than biomass, and its ability to be pumped and stored makes it more economical to transport than solid biomass.
- Because bio-oil can easily be transported, pyrolysis and reforming can be carried out at different locations to improve the economics and for greater convenience.

Fig. 4.21 Schematic of bio-oil distributed reforming approach (Modified from Turner et al. [78])

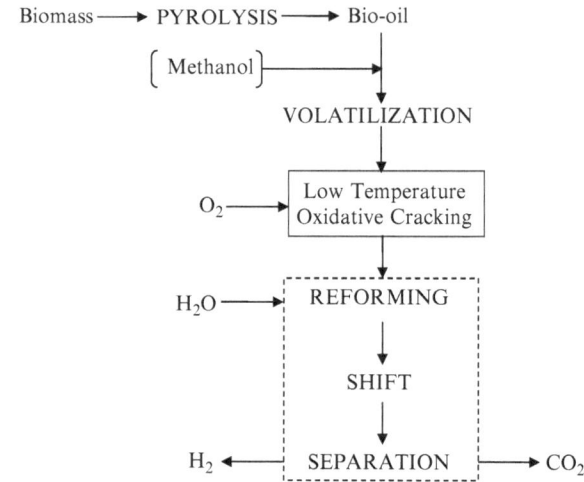

For example, pyrolysis units can be constructed where low-cost feed stocks are available. The bio-oil can be transported to a central reforming plant located at a site with hydrogen storage and distribution infrastructure.

- Hydrogen can also be produced from bio-oil or its fractions at distributed production units, for example, in fueling stations.

The chemical reaction for the pyrolysis and reforming of biomass is as follows.

$$Biomass \rightarrow Bio\text{-}oil + Char + Gas$$

$$CH_{1.46}O_{0.67} \rightarrow 0.71CH_{1.98}O_{0.76} + 0.21CH_{0.1}O_{0.15} + 0.08CH_{0.44}O_{1.23} \quad (4.13)$$

$$CH_{1.98}O_{0.76} + 1.24H_2O \rightarrow CO_2 + 2.23H_2 \quad (4.14)$$

Figure 4.21 shows the aforementioned pyrolysis and reforming of biomass for hydrogen production. Another route of hydrogen production from biomass via thermal cracking of hydrocarbons produced from bio-oil is shown in Fig. 4.22. Biomass/cellulosic and starchy materials on pyrolysis or destructive distillation (i.e., heated in the absence of air) produce wood charcoal as residue and some vapors. The vapors on cooling become liquid and uncondensed gases. The liquid, also known as bio-oil, is a mixture of methanol, acetic acid, acetone, and minor quantities of formic acid and formaldehyde, etc. Acidity is neutralized by treatment with lime soda, and the resulting solution is subjected to distillation. The distillate is a mixture of methanol and acetone that is rich in hydrogen. Hydrogen can be produced from methyl alcohol and acetone via hydrocarbon formation and thermal cracking. The uncondensed residue gases may contain lower hydrocarbons. Therefore, some hydrogen can also be produced from the residue gases.

The biological hydrogen production via biomass (Fig. 4.23) uses the fermentation process. Biomass is rich in carbohydrates, for example, simple sugars (glucose, sucrose, raffinose, etc.) and polysaccharides (starch, celluloses, etc.).

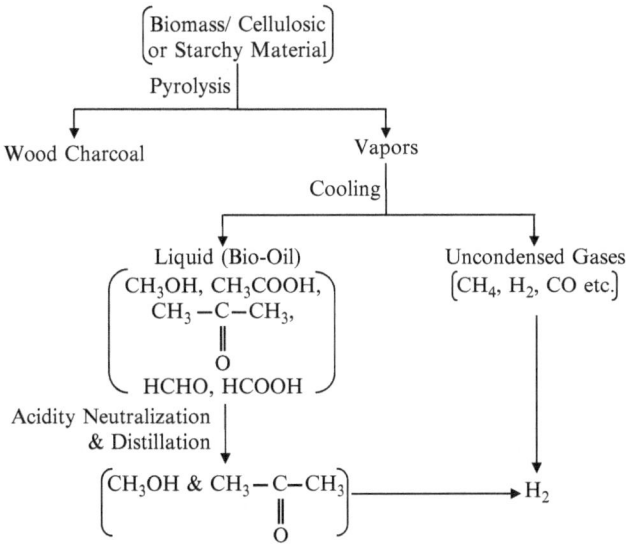

Fig. 4.22 Hydrogen production via pyrolysis of bio-oil

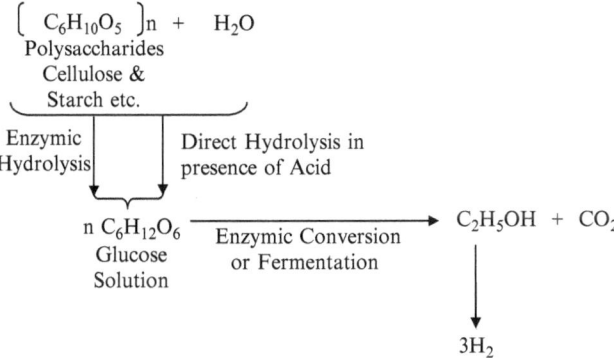

Fig. 4.23 Hydrogen via chemical route and fermentation

All these sugars except glucose are converted into glucose solution in the presence of suitable enzymes. These enzymes include invertase, amylase, maltase, cellulase, and zymase. They possess specific actions: cellulase converts cellulose into glucose, invertase converts sucrose into glucose and fructose, amylase converts starch into glucose, and maltase converts maltose into glucose. Finally, zymase converts glucose into ethyl alcohol. The enzymic conversion is also known as fermentation. Further decomposition of ethanol via hydrocarbon formation and cracking can produce hydrogen. The chemical route of hydrogen production from biomass as shown in Fig. 4.19 is also elaborated in Fig. 4.23. The polysaccharides on hydrolysis in presence of acid produce glucose, which on fermentation gives ethyl alcohol. Hydrogen can be produced by decomposition of ethanol, as already stated. In the chemical route, the heat requirement can be fulfilled by suitable solar technology. The chemical route is less time consuming compared to the biological route as the latter requires more time for the microbiological conversion (fermentation) process.

4.2.10   High-Temperature Electrolysis of Steam

High-temperature electrolysis is a process by virtue of which a significant improvement in electrolytic production of hydrogen can be achieved. Hydrogen production via low-temperature electrolysis mainly depends on the electrolytic conditions of the aqueous media such as alkaline or acidic solutions. Mostly, the electricity is used as the energy input to the cells to split water. However, by raising the temperature of the cell, an increasing amount for decomposition of water can be achieved by thermal energy. The electronic equation for high-temperature electrolysis (HTE) at each electrode is as follows:

Cathodic reaction:

$$2H_2 + 4e^- \rightarrow 2H_2(g) + 2O_2^{2-} \tag{4.15}$$

Anodic reaction:

$$2O_2^{2-} \rightarrow 2O_2(g) + 4e^- \tag{4.16}$$

At very high temperatures, where water exists as steam, the conduction must occur by ionic means rather than by electrolytic as in aqueous condensed systems. Ionic conductors consist of a porous metal oxide ceramic, usually zirconium oxide (ZrO_2) stabilized with yttrium oxide (Y_2O_3) [81, 82]. Electrodes are placed against a thin slab of gas-tight ZrO_2-(Y_2O_3). When steam is supplied to the ionic conductor and a direct current electrical field is applied across the two electrodes, hydrogen forms at the cathode while oxygen ions migrate through the ionic conductor and form oxygen molecules at the anode. The ionic conductors can be arrayed in various geometric configurations (such as in tubular form) to produce a single cell. The cells are further joined either in parallel or series circuitry to form a battery of cells. The battery of cells can further be connected in a bank of batteries to form a large-scale reactor. Many HTE process designs have been developed based on coal as the primary energy source [83]. In the schematic of a lower-cost system (Fig. 4.24), electricity and steam feeds from a coal-fired steam power station and coal-fired steam heater. Ceramic conductor tubes are used to pre-heat the saturated steam from a power plant by a coal-fired heater before being fed to the HTE plant. High-temperature electrolysis occurs and steam is electrolyzed with a temperature drop of 200 °C (360 °F).

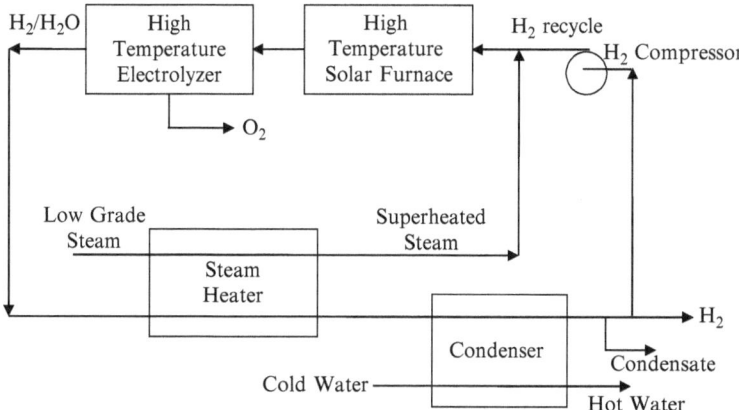

Fig. 4.24 High-temperature electrolysis (HTE) using solar energy (Modified from Steinberg and Cheng [84])

The high-temperature electrolysis process can be designed to convert 50–90 % of the steam to hydrogen and oxygen with an overall thermal efficiency of 35–38 %, based on an efficiency of 35 % electrical power generation from coal [84]. Current densities can vary between 3,000 and 5,000 A/M^3 with cell potentials varying between 1 and 1.6 V for increased hydrogen productivity of the cell.

4.2.11 Solar Thermal Hydrogen Production Via Electricity Generation and Electrolysis

In the schematic of a solar thermal system (Fig. 4.25), the system comprises four main units: concentrating collector, heat engine, generator, and electrolyzer. A thermal heat storage unit can also be added to the circuit to ensure the continuous supply of thermal energy to the heat engine. When solar radiation strikes the concentrating collector, it is concentrated on the absorber surface to which the heat engine is attached. The heat engine uses part of this thermal energy and converts it to mechanical (shaft) work. The remaining thermal energy escapes into the atmosphere and hence is called heat loss. The shaft work is then utilized to generate electricity by using an electrical generator. The electricity generated is then supplied to an electrolyzer unit that electrolyses water into hydrogen and oxygen. The system involves moving parts, for example, heat engine and generator, and therefore requires more maintenance as compared to other systems (e.g., photovoltaic hydrogen system).

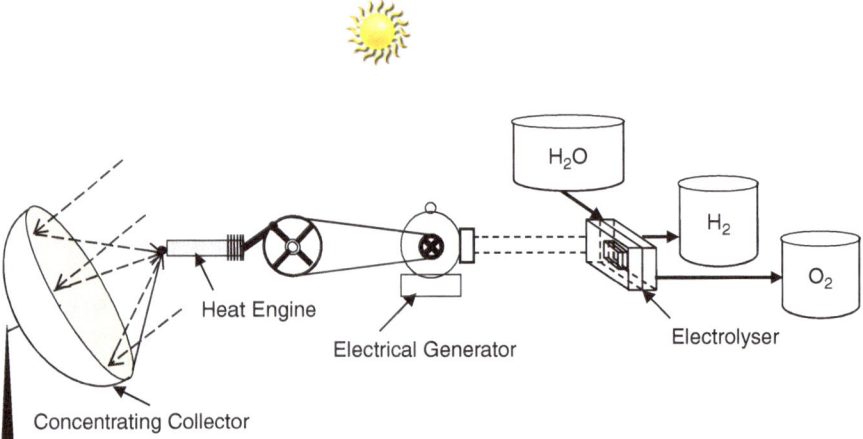

Fig. 4.25 Schematic of a solar thermal hydrogen production system

4.3 Solar Light-Driven Hydrogen Production Methods

Water as the most available source on the Earth is the major resource of hydrogen. Various procedures may be utilized to extract hydrogen, including electrolysis, photolysis, and water purification. Hydrogen production from water splitting requires two molecules of water to donate four-valence electrons to the oxygen nucleus and protons in a general reaction according to $2H_2O \rightarrow 2H_2 + O_2$. This process consumes at least 4.92 eV of energy to generate one molecule of oxygen and two molecules of hydrogen. Additionally, a hydrogen separation method should be utilized to distinguish a pure hydrogen and oxygen stream [85, 86].

Among various water-splitting technologies, water electrolysis is more fully developed. Water splitting via sunlight to produce hydrogen can be achieved through several conversion routes (Fig. 4.26). However, more progress with regard to efficiency improvement and water impurities is required. The energy required for water splitting can be supplied through thermal, electrical, photonic, and biological sources plus hybridization of these possibilities [87].

High-temperature direct thermo-electrolysis of water (> 2,500 K) [89] and lower-grade thermal energy thermochemical methods (<1,000 K) are mainly developed to split the water molecule and separate the products by conducting a series of chemical reactions through intermediate compounds [90]. Photonic radiation can be used in several routes to split the water molecule, either through photolysis or in a hybrid manner such as photo-electrochemical processes.

As shown in Fig. 4.26, the so-called solar thermochemical (STCH) approach is triggered by a photon-to-thermal conversion step followed by a thermal-to-chemical process. In another two-step route, a photon-to-electric step initiates the process and an electric-to-chemical conversion step follows. The concentrating solar thermal (CST) electrolysis pathway comprises three steps of photon-to-heat,

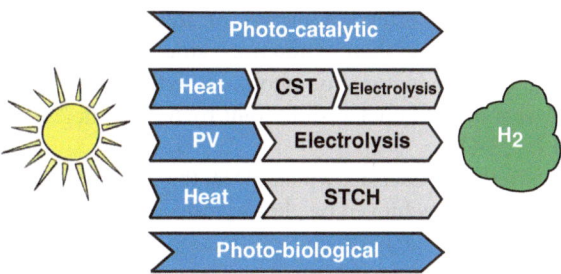

Fig. 4.26 Technical routes for solar-to-hydrogen conversion. *CST* concentrating solar thermal, *STCH* solar thermochemical (Modified from Baniasadi et al. [88])

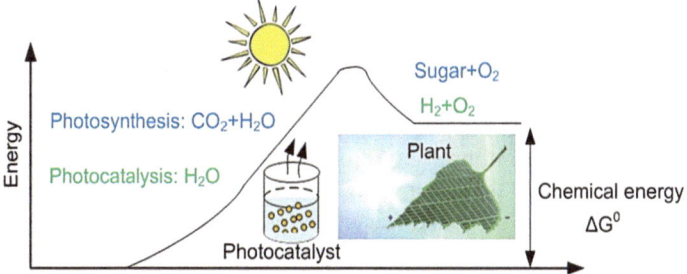

Fig. 4.27 Analogy of photo-catalytic water splitting (artificial photosynthesis) with photosynthesis of green plants (Modified from Arachchige et al. [87])

heat-to-electricity, and electric-to-chemical conversions. The direct process of photon-to-chemical conversion, so-called photo-electrochemical, is another alternative pathway.

Hydrogen and electricity play key roles as effective energy carriers in the future of energy economics. Therefore, the hydrogen production routes such as PV electrolysis and CST electrolysis that implement both could be more beneficial. Recent progress in science indicates the viability of producing inexpensive electricity and hydrogen using semiconductors in the near future. The high energy losses associated with the multistep conversion pathways of the aforementioned techniques have to be treated in some way. From this perspective, the single-stage techniques such as photo-catalysis could be beneficial in terms of minimum process irreversibility (Fig. 4.27). The direct conversion process for hydrogen production, as one of the most promising alternatives, mimics photosynthesis to absorb light and convert water into H_2 and O_2 using inorganic semiconductors through the water-splitting reaction [91].

Many researchers [e.g., [92, 93, 94, 95, 96], have tried to produce hydrogen by different approaches. The pioneering efforts by Fujishima and Honda [92] in 1972 on semiconductors that are capable of absorbing light energy introduced this type of materials as effective means of the water-splitting reaction for hydrogen generation. Since then, numerous studies inspired development of various types of semiconductors for photo-catalytic water dissociation and to enhance their performance under visible light by more efficient utilization of the solar spectrum [93, 94].

Biological hydrogen production, which offers the possibility of being renewable and carbon neutral, can be achieved by photosynthesis, fermentation, and microbial electrolysis cells. Biological hydrogen production has several advantages over hydrogen production by photo-electrochemical or thermochemical processes. Biological hydrogen production by photosynthetic microorganisms, for example, requires the use of a simple solar reactor such as a transparent closed box, with low energy requirements.

4.3.1 Photo-Electrochemistry of Water Splitting

The possibility of transmitting energy to every point of a volume using photonic radiation is applicable to the water cleavage process. Because pure water does not absorb radiation in the visible and near-ultraviolet ranges, the dissociation of water is technically possible by either electromagnetic rays, for example, exposing water to higher-frequency radiation such as extreme UV or X-and gamma rays, or using molecular photosensitizers dissolved/suspended in water, to capture solar radiation in the visible and UV ranges. The latter approach constitutes important paths for solar-driven water splitting. In this regard, the interaction of photonic radiation with a photo-catalyst and water molecules is important for technology development.

The theoretical potential for water splitting process is 1.23 eV per molecule. This energy corresponds to the wavelength of 1,010 nm that makes about 70 % of the solar-irradiated photons eligible for driving the water cleavage reaction. However, the amount of the energy per photon should be higher than the minimum value of 1.23 eV to compensate the intrinsic energy losses associated with the redox reactions on the surface of the photo-catalyst [95].

Photo-catalytic water splitting is initiated with the absorption of light photons with energies higher than the band-gap energy (Eg) of a semiconductor. As depicted in Fig. 4.28, following the absorption process excited photoelectrons are generated in the conduction band (CB) and holes in the valence band (VB) of the semiconductor. Once the photo-excited electron–hole pairs are created in the material bulk, they migrate to the surface separately (represented by routes "i" and "ii").

This process is in continuous competition with the electron–hole recombination process (route "iii" in Fig. 4.28) that leads to heat generation. Eventually, the water molecules in the vicinity of catalytically active sites are reduced and oxidized by photo-induced electrons and holes to produce gaseous hydrogen and oxygen, respectively. The decomposition of water into H_2 and O_2 requires a large positive

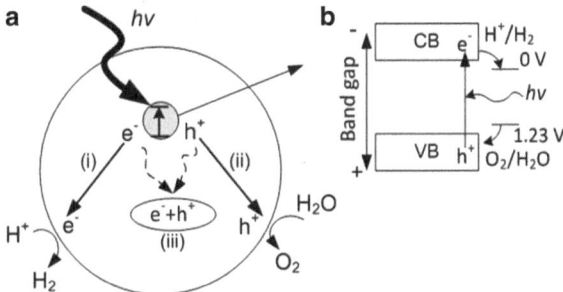

Fig. 4.28 Basics of photo-catalytic water splitting. (**a**) Photo-excitation and subsequent processes on catalyst surface. (**b**) Photo-excitation and electron–hole generation. *CB* conduction band, *VB* valence band (Modified from Naterer et al. [95])

change in the Gibbs free energy ($\Delta G_0 = 237$ kJ mol^{-1}, 2.46 eV per molecule). The following reactions are performed in an acidic medium.

Oxidation:

$$H_2O + 2h^+ \rightarrow 2H^+ + 1/2O_2 \tag{4.17}$$

Reduction:

$$2H^+ \; + \; 2e^- \rightarrow H_2 \tag{4.18}$$

The water dissociation process refers to either half-processes such as water reduction to hydrogen and hydroxyl ions or water oxidation to oxygen and protons, or to the complete conversion of water to hydrogen and oxygen. The photochemical process of water splitting does not necessarily involve a complete water-splitting process. Photo-induced half-reactions to generate hydrogen mainly occur according to $\left[2H_2O + 2e^- \xrightarrow{h\nu} H_2(g) + 2OH^- \right]$ in alkaline medium. The hydroxide ion by-product can be further processed through electro-catalytic oxidation to generate oxygen.

Two major configurations of photo-catalytic technique for water cleavage are electrode based and particle based. The first method implements two electrodes immersed in an aqueous electrolyte: one is the semiconductor photo-catalyst exposed to light and the other is a counter-electrode to facilitate the electron flow circuit (Fig. 4.29). The second system is based on photo-catalysts in the form of suspended particles in an electrolyte (Fig. 4.29).

In this system, each particle acts as a micro-photo-electrode to conduct redox active sites for splitting water. In comparison, particle-based configuration has some disadvantages with respect to electron–hole separation efficiency and reverse reaction of the products on the surface of the photo-catalyst that leads to conversion rates. However, photo-catalysts in particle form have the advantage of much

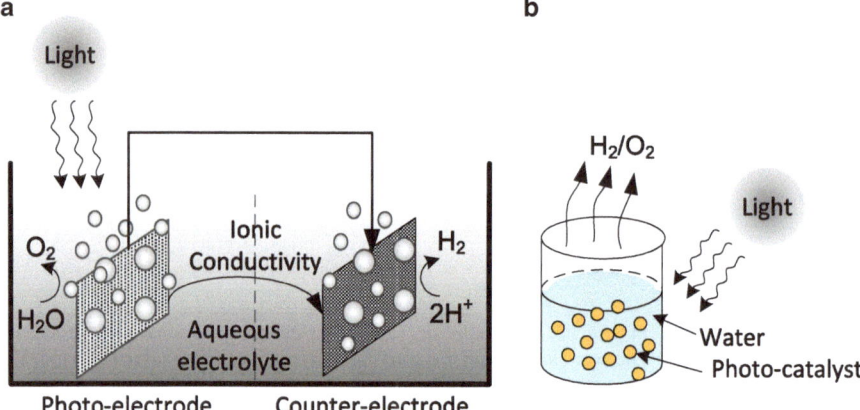

Fig. 4.29 General configurations for photo-catalytic water splitting: electrode based (**a**) and particle based (**b**) (Modified from Baniasadi et al. [88])

simpler and less expensive synthesis. Aside from these two configurations that undergo heterogeneous catalysis, another approach of homogenous catalysis based on interaction of complex molecular structures in a solution is also feasible.

4.3.2 Heterogeneous Photo-Catalysis

Photo-catalysis in solid phase implemented in reaction of water dissociation in the liquid phase is denoted as heterogeneous. These catalysts are applied in powder and electrode form. Photo-electrodes are being widely developed to enhance the efficiency of water catalysis. The photo-electrode structure consists of a conductive material such as noble metals doped with photo-catalysts. The conductivity of the electrode material facilitates the migration of electrons from its conduction band to the valence band and eventually to the active catalytic sites where water electrolysis is performed. The photo-electrodes in a photo-reactor configuration are exposed to solar radiation through transparent windows. Direct incident light together with electrical potential bias, supplied by an external electrical power supply, are utilized to activate active sites.

The small band gap between the valence and conduction bands of semi-conductors make them suitable to develop heterogeneous photo-catalysis. Some semiconductors such as titanium dioxide have both photosensitization and photo-catalytic attributes. Thus, when the semiconductor is exposed to light, it absorbs photons and dislocates electrons from the valence band to the conduction band. It eventually facilitates redox reactions at the photo-catalyst surface. The performance of photo-catalysts is usually enhanced to overcome the redox potentials through two possible routes:

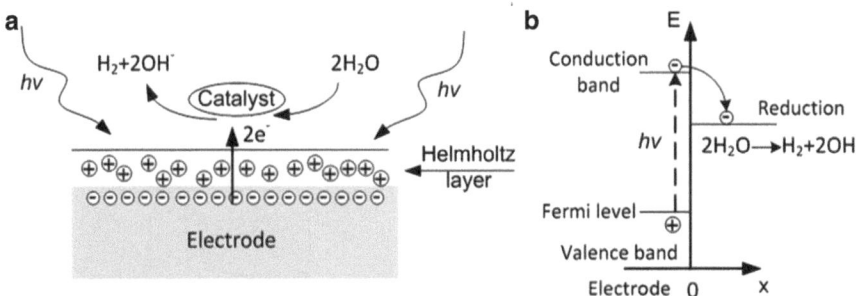

Fig. 4.30 (**a**) Heterogeneous photo-catalysis mechanism. (**b**) Reduction reaction (Modified from Naterer et al. [95])

i. Applying a bias voltage to the electrode via an external power supply; or
ii. Electrode coating or doping with photo-synthesizer materials.

The heterogeneous photo-catalysis process is illustrated in Fig. 4.30: the reaction mechanism is introduced in Fig. 4.30a and the reduction pathway is shown in Fig. 4.30b. Following the light absorption and activation of catalytic sites, heterogeneous catalysis occurs at the solid surface. The catalyst reduces the activation energy without being consumed. To quantify the ability of the photo-catalyst to interact with reactant molecules, the turnover number (TON) is defined as the number of reactants that are treated with catalyst before degradation. The turnover frequency (TOF) is also an indicative parameter that represents the number of converted reactants by photo-catalyst per unit of time.

A layer of ionized solvent molecules called the Helmholtz layer hosts the electron transfer from the active site to the reactant molecules. This layer is adjacent to the solid–liquid interface with a thickness of 1 Å. The electrons migrate to the reacting molecules through the Helmholtz layer. The semiconductor has the same Fermi energy levels as the Helmholtz layer; therefore, the process of electron transfer is iso-energetic. The dislocated electron in the conduction band then falls into a lower energetic level to accomplish the reduction reaction (Fig. 4.30a).

Figure 4.31 shows the band levels of various semiconductor materials. The pH of the solution usually changes the band levels by -0.059 V/pH for oxide materials. According to the information in Fig. 4.31, ZrO_2, $KTaO_3$, $SrTiO_3$, and TiO_2 have suitable band structures for water splitting. These materials are usually modified with co-catalysts to be active for water splitting.

The n-type semiconductors such as cadmium sulfide or zinc sulfide are recognized as potential photo-catalysts for large-scale hydrogen production. However, they involve anodic photo-corrosion in aqueous solutions that leads to the formation of sulfur or sulfate ions according to

$$CdS + 2h^+ \rightarrow cd^{2+} + S \tag{4.19}$$

However, they are promising photo-catalysts for H_2 evolution under visible and ultraviolet light irradiation if reducing agents acting as hole scavengers, such as

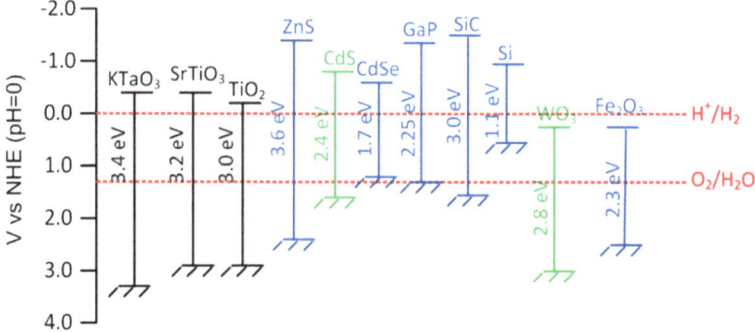

Fig. 4.31 Relationship between band structure of semiconductor and redox potentials of water splitting (Modified from Ni et al. [97])

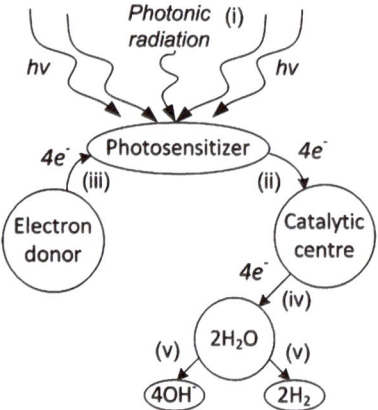

Fig. 4.32 Homogeneous photo-catalysis process with complex molecular structure (Modified from Naterer et al. [95])

S^{2-}, SO_3^{2-}, or $S_2O_3^{2-}$, exist in the aqueous solution to stabilize cadmium sulfide and zinc sulfide efficiently [96].

4.3.3   Homogeneous Photo-Catalysis

Homogeneous photo-catalysts consist of complex molecular structures that perform photo-catalytic water splitting starting from photosensitization. Subsequently, charge separation, charge transfer, and catalysis processes are conducted by the molecular structure. Figure 4.32 shows the process of homogeneous photo-catalysis for hydrogen evolution.

The series of events initiates with photon absorption of the photo-sensitizer that turns it to its excited state. The excited-state form of the photo-sensitizer interacts with the catalyst and translocates an electron to the active catalytic center. The photo-sensitizer returns to reactive state by absorbing an electron from the electron donor dissolved in the solution. The electron donor remains stable in the solution. The water reduction reaction is then accomplished by electron transfer through catalytic active centers. Four catalytic cycles are performed following four photon absorptions to complete a water dissociation reaction.

4.3.4   Photo-Catalytic Water Splitting

The development of an energy scenario based on the use of two natural resources, water and sunlight, is a challenging solution to the energy crisis of the world. Conversion of solar energy into chemical energy through light-driven water splitting generates the environmentally benign gas oxygen and hydrogen, a carbon-free fuel with the highest energy output relative to its molecular weight. Moreover, this approach provides an attractive solution for storing the tremendous amount of sunlight energy falling on the Earth. In that context, natural photosynthesis is a great source of interest for the scientific community. Conversion of light to chemical energy is indeed achieved by photosynthetic organisms. As far as the energetic aspect is concerned, the photosynthetic process is a fascinating example of efficiency [98, 99].

Methods derived from natural photosynthesis are therefore highly attractive for the development of novel hydrogen production technologies. Understanding this biological process and exploiting this knowledge for designing original synthetic molecular systems achieving light-to-chemical energy conversion is the basis of a large field of research called artificial photosynthesis [100, 101]. Molecular light-harvesting arrays have been developed to mimic the antenna effect, which is collecting energy of many photons to transfer it directionally to a final acceptor achieving charge separation. Dyad and triad models based on the association of a photo-sensitizer to either an electron donor, an electron acceptor, or both components, have been designed to reproduce light-induced spatial charge separation. These first two topics have been the core of artificial photosynthesis, and also the subject of comprehensive reviews [102–107].

Currently, research is often focused on the independent development of the two half-reactions, through the combination of a photo-sensitizer with a suitable catalyst for either the oxidation or the reduction of water, together with a sacrificial electron acceptor or donor, respectively (Fig. 4.33). Efforts have also been devoted toward the synthesis of single-component photo-catalysts, either by supramolecular assembly or by covalent linking of the light-harvesting unit to the catalyst. This design can improve electron transfer between these two units, which is relevant to the spatially controlled assembly of the various photosynthetic components in the membrane that is largely responsible for the efficiency of the natural photosynthetic

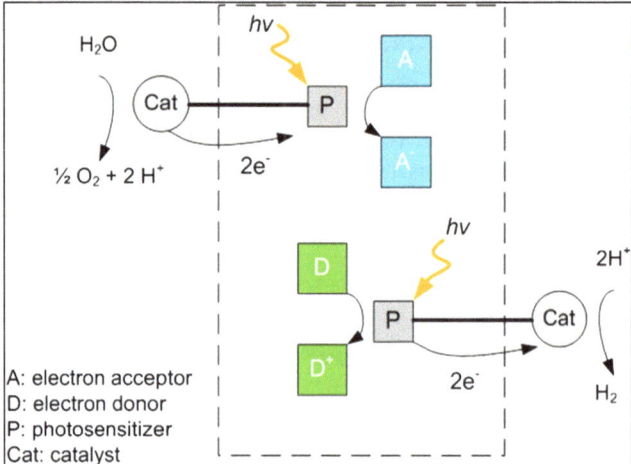

Fig. 4.33 Schematic representation of homogeneous multi-electron photo-catalysis for light-driven water splitting (Modified from Baniasadi et al. [88])

process. Supramolecular complexes in a photo-catalytic hydrogen production scheme that result in high turnover rates and numbers are an area of interest [108]. Supramolecular solar H_2 photo-catalysts usually consist of Ru or a light-absorptive metal, a light absorber (LA), coupled to a metal-based catalyst including Pt, Pd, Rh, Co, bio-inspired di-iron dithiolates, or single-component dirhodium or Pt systems [109–111].

Encouraging results have been obtained by Fihri et al. [112, 113], Li et al. [114], and Brauns et al. [115] with cobalt-based supramolecular photo-catalysts, including supramolecular photo-catalysts for H_2 production based on cobaloxime centers. Some series of ruthenium-based photo-catalytic assemblies have established that a conjugated bridge is not required for the activity. The long-lived metal-to-ligand charge-transfer (MLCT) excited state of $[Ru(bpy)_3]^{2+}$ has motivated photochemical and photophysical studies leading to light-to-energy conversion processes [116]. The MLCT excited state of $[Ru(bpy)_3]^{2+}$ and its analogues have the required energy to split water into hydrogen and oxygen, but it requires complicated multi-component systems for operation.

Recently, progress has been achieved in H_2 production using homogeneous molecular catalysts powered by the sensitization of a molecular light absorber (LA) [117–120]. More recently, multi-component systems have been identified that use Ir LAs with Rh catalysts or a Pt LA with a co-catalyst of 1,000 t in 10 h [121–123].

Weber and Dignam [124] have developed photochemical water-splitting catalysts using supramolecular devices that are able to capture the incident solar radiation and generate electrons or holes at the active center where water reduction or oxidization occurs, respectively. Such systems mimic natural photosynthesis

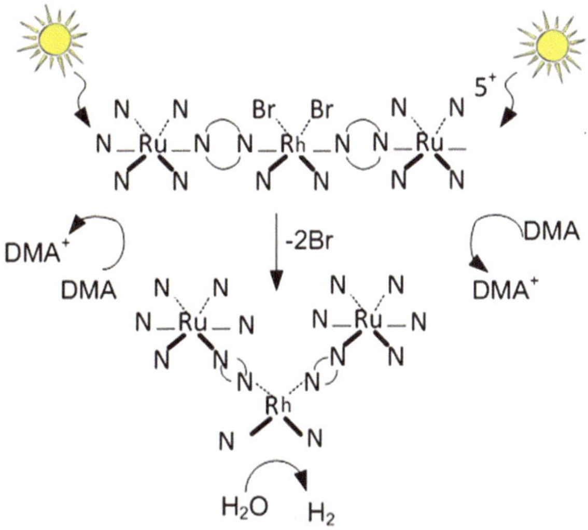

Fig. 4.34 Photochemical scheme for H_2 fuel production from H_2O using a $[\{(bpy)_2Ru(dpp)\}_2RhBr_2]^{5+}$ photo-initiated electron collector (Modified from Weber and Dignam [124])

and mainly consist of supramolecular complexes of organic molecules that possess active metallic centers. Photo-initiated electron collection affords Rh species that provide reducing equivalents to H_2O to make H_2 fuel (Fig. 4.34). They have reported a photo-catalytic system for H_2 production using the $\left[\left\{(bpy)_2Ru(dpp)_2RhBr_2\right\}\right]^{5+}$ photo-catalyst, DMA, H_2O, and DMF solvent that provides a catalyst with a turnover number of 280 in 19.5 h and a maximum quantum efficiency of $\varnothing = 0.023$.

4.3.5 Photo-Electrochemical Water Splitting

The free energy change for the conversion of one molecule of H_2O to H_2 and $\frac{1}{2}$ O_2 under standard conditions is $\Delta G = 237.2$ kj/mol, which, according to the Nernst equation, corresponds to $\Delta E^\circ = 1.23$ V per electron transferred. For photochemical water reduction to occur, the flat-band potential of the semiconductor (for highly doped semiconductors, this equals the bottom of the conductance band) must exceed the oxidation potential of water of $+1.23$ V versus NHE at pH $= 0$ ($+0.82$ V at pH $= 7$) (Fig. 4.35a). A single band gap device requires, at a minimum, a semiconductor with a 1.6–1.7 eV band gap to generate the open circuit potential required to split water. Once other voltage-loss mechanisms (i.e., catalysis) are accounted for, a band gap above 2 eV is generally necessary [125].

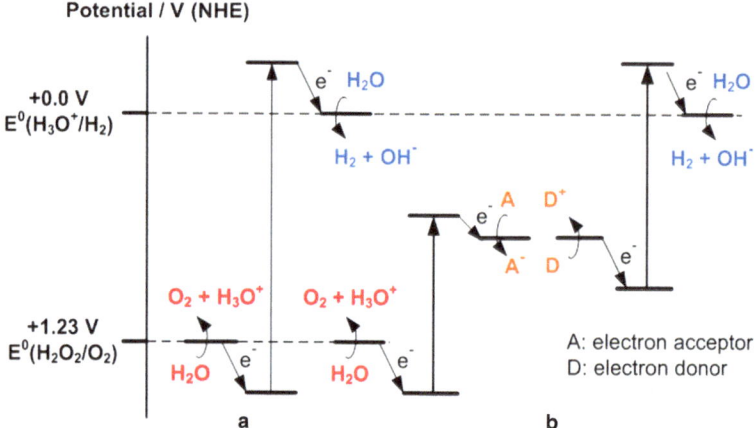

Fig. 4.35 Diagram of required potential for photochemical water splitting at pH = 0: single semiconductor system (**a**) and dual semiconductor system (z scheme) using a redox shuttle (**b**) (Modified from Kato et al. [127])

The smallest band gap achieved so far in a functional catalyst is 2.30 eV in NiO/RuO$_2$ − Ni:InTaO$_4$ [126]. Semiconductors with smaller band gaps or lower flat-band potentials require a bias voltage or external redox reagents to drive the reaction. Alternatively, two or more small band gap semiconductors can be combined to drive water oxidation/reduction processes separately via multi-photon processes (Fig. 4.35b) [127].

The water-splitting reaction to hydrogen and oxygen can be facilitated by many inorganic semiconductors, such as TiO$_2$, which was discovered in 1971 by Fujishima and Honda [92]. During the past decade, more than 130 materials and derivatives have been developed for overall water catalysis to hydrogen and oxygen or water oxidation or reduction by employing external redox agents. The best compounds in terms of quantum efficiencies (η_ϕ) are NiO-modified La/KTaO$_3$ ($\eta_\phi = 56$ %, pure water, UV light) [127], ZnS ($\eta_\phi = 90$ %, aqueous Na$_2$S/ Na$_2$SO$_3$, light with $\lambda > 300$ nm) [128], and Cr/Rh-modified GaN/ZnO ($\eta_\phi = 2.5$ %, overall pure water splitting, visible light) [129, 130]. Up to now, no material has been developed to catalyze water splitting with visible light at quantum efficiency greater than 10 %. Here, 10 % is the goal for commercial applications [131].

The use of two semiconductor materials remains an attractive option for capturing a large portion of the solar spectrum, with the two band gaps tuned to absorb complementary portions of the solar spectrum [125]. A dual band gap cell configuration consists of an n-type material (photo-anode) with a p-type semiconductor (photo-cathode) in series. The n-type material is capable of water oxidation with sufficiently positive valence band and reasonable photocurrent, and the p-type material has a smaller band gap that would drive the hydrogen evolution reaction.

Fig. 4.36 Sketch of the dual cell for photo-catalytic water splitting, with separate H_2 and O_2 evolution (Modified from Selli et al. [135])

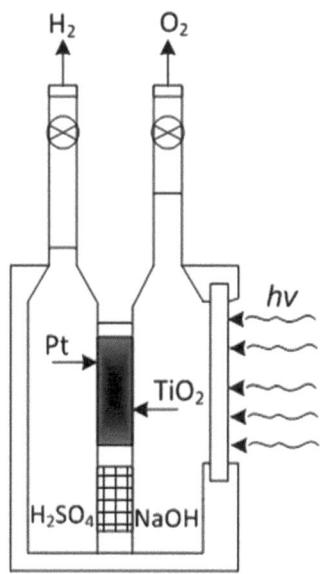

4.3.6 Photo-Catalytic Overall Water Splitting

Development of new types of photo-catalysts for water splitting to produce hydrogen by visible incident light has been extensively studied in the past decade [132, 133]. Traditionally, water splitting with photo-catalysts in powder form produces a mixture of H_2 and O_2. Therefore, efficient devices must be designed on the basis of the use of heterogeneous or homogeneous photo-catalysts to separate H_2 and O_2 in water photo-splitting processes [134].

Selli et al. [135] have developed a photo-catalytic water-splitting process by the use of a new two-compartment Plexiglas cell. This design provides an effective decomposition of H_2O into H_2 and O_2, where oxygen evolves in a separate half-cell by an illuminating photo-active Ti electrode (Fig. 4.36). Hydrogen production under irradiation with a UV lamp at a wavelength above 300 nm was performed without use of any sacrificial reagent. The photonic efficiency is reported as 2.1 %, whereas it reduces to 0.36 % under irradiation with a visible wavelength lamp ($\lambda > 350$ nm) [65, 135].

Sun et al. [136] have designed a photo-electrochemical device by an assembly of a molecular Ru catalyst with pH-modified nafion on a dye-sensitized nanostructured TiO_2 film as the anode to split water into O_2 and H_2 under visible light. A Pt foil has been used as the cathode. A small bias of -0.325 V versus Ag/AgCl is required to boost the conduction band of TiO_2 for direct reduction of protons to hydrogen and perform the complete water splitting. The strong acidity of commercial nafion is found as the reason for an increase of over-potential associated with water oxidation and the rapid decay of the photocurrent.

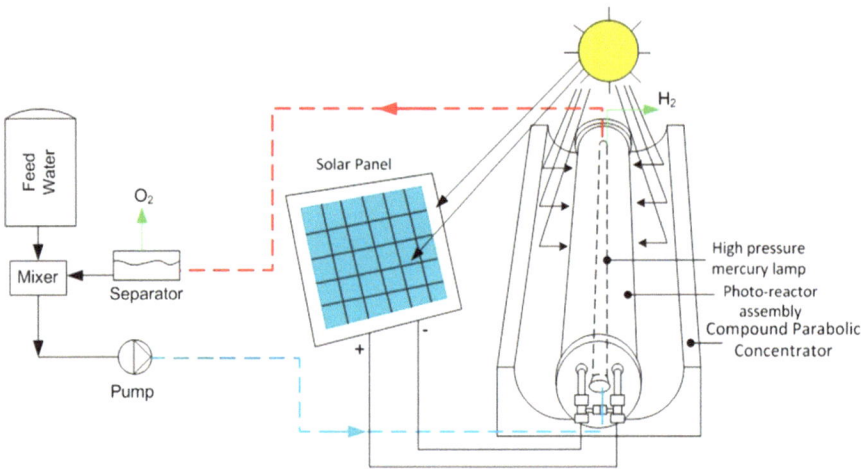

Fig. 4.37 Pilot plant-scale hybrid photo-catalytic system for continuous operation, under sunlight and artificial illumination (Modified from Baniasadi et al. [138]

A new hybrid system for hydrogen production via solar energy was developed and analyzed by Baniasadi et al. [137]. To decompose water into hydrogen and oxygen without the net consumption of additional reactants, a steady stream of reacting materials must be maintained in consecutive reaction processes, to avoid reactant replenishment or additional energy input to facilitate the reaction. The system comprises two reactors, which are connected through a proton-conducting membrane. Oxidative and reductive quenching pathways are developed for the water reduction and oxidation reactions. Supramolecular complexes $[\{(bpy)_2Ru (dpp)\}_22RhBr_2](PF_6)_5$ are employed as the photo-catalysts, and an external electric power supply is used to enhance the photochemical reaction. The energy and exergy efficiencies at a system level are analyzed and discussed. The maximum energy conversion of the system is improved up to 14 % by incorporating design modifications that yield a corresponding 25 % improvement in the exergy efficiency.

A photo-catalytic reactor activated by sunlight and UV-visible lamps capable of continuous operation under real process conditions was examined by Baniasadi et al. [138] for scale-up purposes. Figure 4.37 shows a schematic of a hybridized photo-catalysis reactor that utilizes light energy from both the sun and a lamp, in conjunction with electrodes, to substitute sacrificial electron donors that consume photo-generated holes and transfer charges to a metallic active center of a catalyst that leads to hydrogen production from water. Solar panels are utilized to generate electricity and deliver electrons through the electrode–solution interface. Utilization of UV-visible lamps inside a sunlight concentrator provides the capability of performing photoreactions during nighttime or cloudy periods.

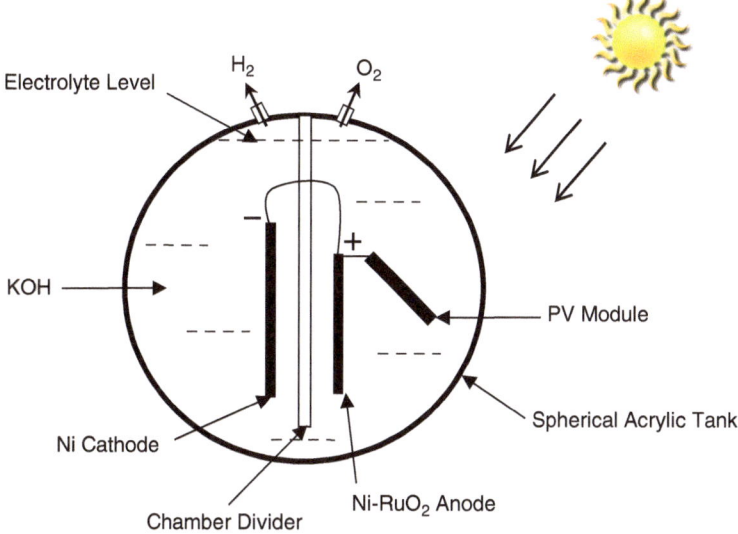

Fig. 4.38 Schematic of a spherical tank reactor. *PV* photovoltaic (Modified from Kelly and Gibson [139])

4.3.7   Photo-Electrolysis

Another method to convert sunlight into hydrogen is by photo-electrolysis of water that uses photo-electrochemical light-collecting systems (PEC, photo-electrochemical cell) to power the electrolysis of water. Recently, Kelly and Gibson [139] conducted some tests on two simple and inexpensive photo-electrochemical (PEC) reactors, namely, teardrop-shaped plastic-film bag reactors and an acrylic spherical tank reactor, capable of producing hydrogen via water electrolysis using solar energy, and found that the latter had the best light-focusing properties, with an increase up to 3.7 in the solar energy (and thus hydrogen production) for PEC cells mounted at the optimal point within the reactor electrolyte. They concluded that by increasing the solar irradiance on the PEC photo-electrode, the hydrogen production increases, and this also can help to reduce the system cost. A schematic of their spherical tank reactor is shown in Fig. 4.38.

The positive and negative terminals of the PV modules were connected to anode and cathode, respectively, which are separated by a chamber divider. The tank was filled with the 5 M KOH electrolytic solution. When sunlight falls on the PV module, it produces electricity, which further negotiates the chemical reaction and produces hydrogen on the cathode and oxygen on the anode side of the system, respectively. The spherical shape of the rector is helpful to focus solar radiation on the PV module by properly orienting the PV module toward the sun or simply by tilting the reactor so the PV cells face the sun, thereby increasing the hydrogen

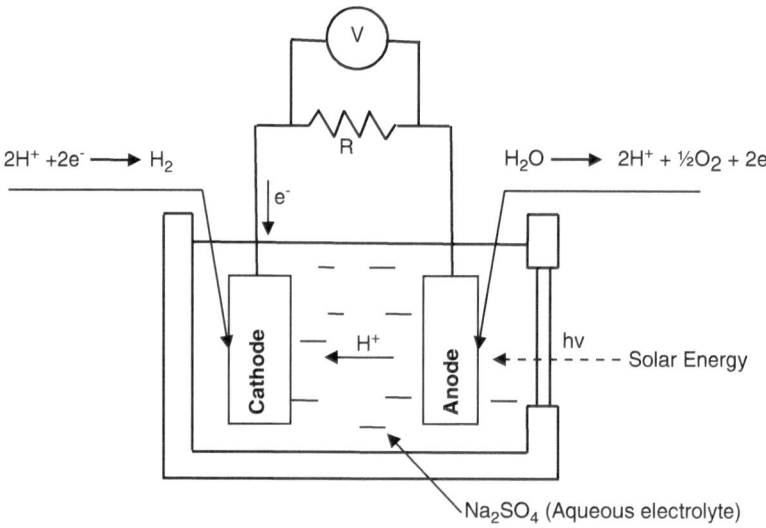

Fig. 4.39 Schematic of solar photo-electrolysis (Modified from Fujishima and Honda [92])

output per unit area. Because the electrolyte solution takes the shape of its curved container, the reactor acted as a convex lens that focused the solar radiation onto the PV cells at an increased intensity. Further, it is possible to focus light onto the PV cell by a combination of light-focusing properties of the reactor together with that of lenses such as Fresnel lenses. Another advantage of having the PV module dipped inside the KOH solution is that the performance of the solar cells in terms of electrical efficiency would not be affected by the thermal energy (heat) of the solar radiation. In other words, the KOH solution would cool down the solar cells if the concentrated solar radiations try to increase the temperature of the solar cells of the PV module.

Fujishima and Honda [92] also proposed a photo-electrochemical hydrogen production system that uses a single PEC (Fig. 4.39). When exposed to sunlight, a semiconductor photo-electrode (anode or cathode), submerged in an aqueous electrolyte, generates sufficient voltage to split water molecules; typically, the other electrode is a metal [140]. Photons with energies greater than the semiconductor band gap can be absorbed by the semiconductor, creating electron–hole pairs that are split by the electric field in the space-charge region between the semiconductor and the electrolyte. The electric field reflects the band bending of the conduction and valence band edges at the semiconductor surface and is necessary to supply the free carriers to the appropriate electrode. The overall cell reaction then becomes [141]:

$$2h\upsilon + H_2O(l) \rightarrow 1/2O_2(g) + H_2(g) \tag{4.20}$$

where h is Planck's constant and υ is the frequency.

Photo-electrolysis integrates solar energy absorption and water electrolysis into a single photo-electrode. The device does not require a separate power generator and electrolyzer [142]. The maximum theoretical efficiency is about 35 % [143]. In current research, it has been seen that photo-electrolysis systems reached first-law thermodynamic efficiency of about 18 % [144]. However, this efficiency was obtained for only a very small amount of hydrogen production. The solar-to-hydrogen conversion efficiency of the materials has not yet met requirements for practical use because of the limitation of the usable solar spectrum. It is expected that technological advancement and speedy commercialization will make such systems feasible and practical. Although photo-electrolysis cells are simple and do not require the complex manufacturing steps necessary in the formation of p/n and n/p photovoltaics, they require large land space and semiconductor requirements. Their lifetime is also extremely short, and some operational/technical details are unknown [145]. Moreover, they suffer from material problems.

4.3.8   Photo-Biological Hydrogen Production

Photo-biological hydrogen production uses the same processes as plant and algal photosynthesis for hydrogen production. The biological hydrogen production can be classified into two types: (i) light-dependent process and (ii) light-independent process. The first process, that is, the light-dependent process, includes direct or indirect bio-photolysis and photo-fermentation whereas the second, the light-independent process, includes dark fermentation [146]. Plant and algal photosynthesis results in the splitting of water into oxygen and a reducing agent strong enough to reduce CO_2, or protons, to carbohydrates or hydrogen, respectively. Photo-biological production of hydrogen from water (bio-photolysis) involves an efficient biological converter, microalgae, and a photo bioreactor. Microalgae are suitable for such processes, as they exhibit hydrogen evolution under some conditions, and they can be cultured in closed systems that can permit hydrogen capture. Microalgal strains must be developed that exhibit high hydrogen production rates and photosynthetic efficiencies in dense cultures at full solar intensities. The photo-bioreactor must expose the H_2-producing cultures to sunlight while allowing recovery of the gas [10]. One advantage of biological processes is that they can be catalyzed by microorganisms in an aqueous environment at ambient temperature and pressure [146]. Biological methods for solar hydrogen production have not yet been developed for commercial use, except for laboratory-stage and small (<10 m^2) outdoor demonstration scale systems [147]. This technology is still under development because of relatively low efficiency of photosynthesis: trees and agricultural crops convert sunlight at efficiencies less than 1 % [148].

The photosynthesis process can be understood as follows. The light and dark reactions are the two main parts of the photosynthesis process by virtue of which plants and trees produce carbohydrates (sugars) and starch, which are essential for their life and growth; in due course, CO_2 is utilized and water is split into oxygen

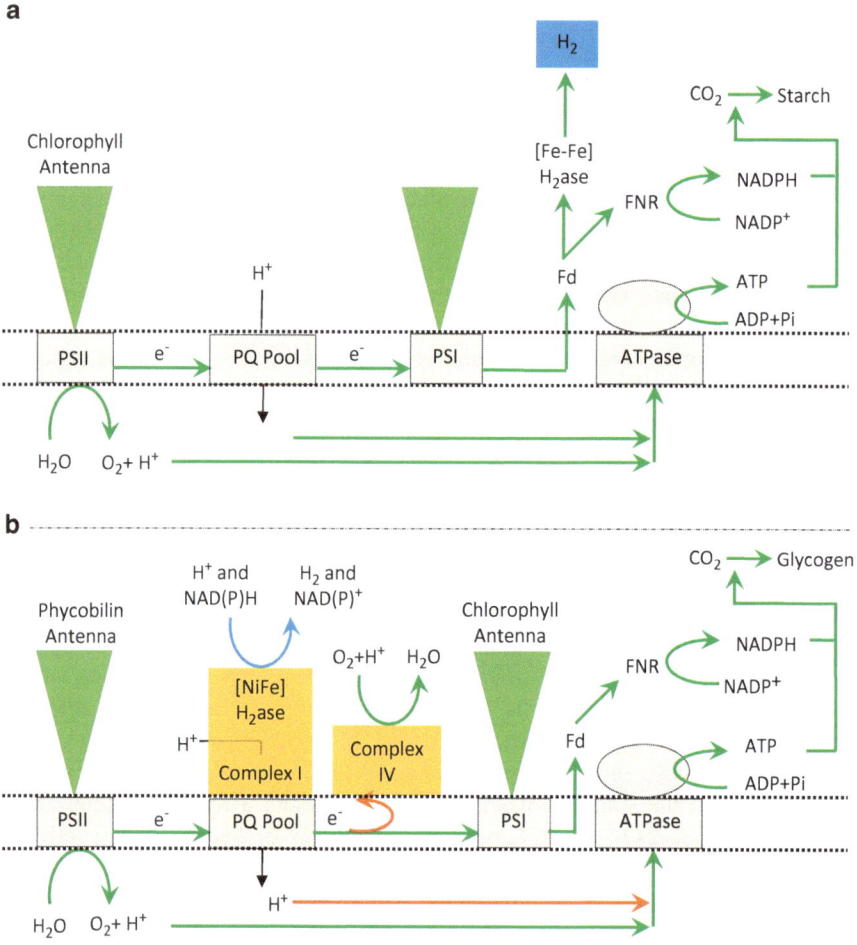

Fig. 4.40 Hydrogen production via the photo-biological process (*PS* photosynthesis) (Modified from Turner et al. [78])

and hydrogen. The light reaction is also known as the oxygenic photosynthetic organism in which the sunlight is absorbed by light-harvesting pigments (chlorophylls, carotenoids, phycobilins), also known as "antenna" chlorophyll molecules in special cell membranes (thylakoids), and then it is transferred very efficiently to membrane-bound "reaction center" chlorophylls. Photo-production of hydrogen by microorganisms is associated with the light absorption and charge separation reactions of photosynthesis.

Figure 4.40a,b shows photosystems II and I, which are connected in series and work together. At the photosystem II (PSII) reaction center, water is oxidized/split into oxygen, protons, and electrons first, thereby releasing oxygen as a by-product to the atmosphere. This reaction takes place via a series of electron transfer reactions (known as the Z-scheme of photosynthesis). The exited electron then delivered through photosystem I (PSI) to a soluble carrier protein, ferredoxin. The electrochemical reactions generate two vital "energy-rich" biological compounds, namely, adenosine triphosphate (ATP) and reduced pyridine nucleotide (NADPH). The electron carriers (e.g., ferredoxin and NADPH) transfer photosynthetically generated reductants to the hydrogen-producing enzymes hydrogenase or nitrogenase instead of diverting them to CO_2 fixation, their normal physiological fate [78]. The hydrogenase-containing organisms such as cyanobacteria and green algae can also extract reductants from water. These organisms produce hydrogen without the input or output of carbon-based molecules. In green algae, ferredoxin transfers photosynthetic reductants to an [FeFe]-hydrogenase where the electrons are recombined with protons that yield hydrogen gas (Fig. 4.40a). In cyanobacteria, the electrons are initially used to reduce $NADP^+$ to NADPH, which interacts with the bidirectional [NiFe]-hydrogenase that is a component of a protein complex which is presented by complex I (Fig. 4.40b). The role of complex I in respiration is to oxidize NADH and transfer electrons to a quinone molecule. In cyanobacteria, this complex seems to be able to oxidize NADPH as well, and to transfer electrons either to a respiration-linked quinone or to a photosynthetic plastoquinone (Fig. 4.40b). The dark reaction involves the formation of carbohydrate (sugars) from carbon dioxide and the products of the light phase, that is, ATP and NADPH, which are used within cells, via a series of biochemical intermediates.

Chapter 5
Thermodynamic Analysis

5.1 Introduction

In this chapter, photovoltaic and solar thermal hydrogen production systems are analyzed. The two different routes are shown in Fig. 5.1. In the photovoltaic (PV) route, DC electricity is generated first by PV panes and then stored in a battery bank. The DC electricity can be converted to AC electricity by using an inverter, and then this electricity is further used to run an electrolyzer. In the solar thermal route, the thermal energy of solar radiation is first collected and concentrated using a concentrating solar collector. A thermal storage system may also be used to ensure the continuous supply of thermal energy. Then, by using a heat engine the thermal energy is converted into mechanical shaft work, and by coupling a generator to the heat engine, the shaft work is converted into electricity. This electricity is used by an electrolyzer for water electrolysis.

5.2 Energy and Exergy Efficiencies of Solar H_2 Production Via Thermochemical Cycles Using Concentrated Solar Energy

Solar hydrogen production via the thermochemical cycle involves a solar concentration system and receiver and reactor systems. For the thermochemical process the required temperature is very high (2,000 K); therefore, heliostat field mirrors are the perfect candidate. Concentrated thermal energy is directed to the receiver and then to a cavity reactor where the thermochemical process takes place. An example of a preferred solar reactor is the fluid wall tubular reactor [15]. Thermal insulation in the form of a glass window, thermal-resistant brick lining, and a fluid jacket are provided around the cavity reactor to minimize thermal losses from the reactor to ambient from radiation, conduction, and convection.

I. Dincer and A.S. Joshi, *Solar Based Hydrogen Production Systems*,
SpringerBriefs in Energy, DOI 10.1007/978-1-4614-7431-9_5, © The Author(s) 2013

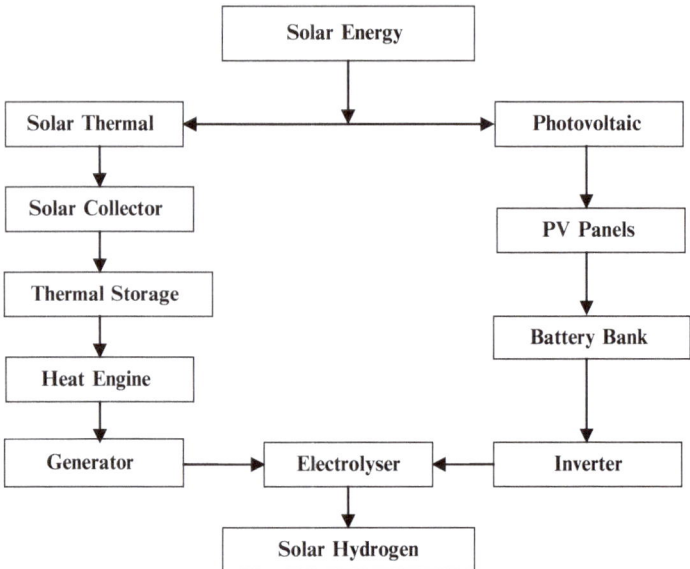

Fig. 5.1 Two routes of solar hydrogen production

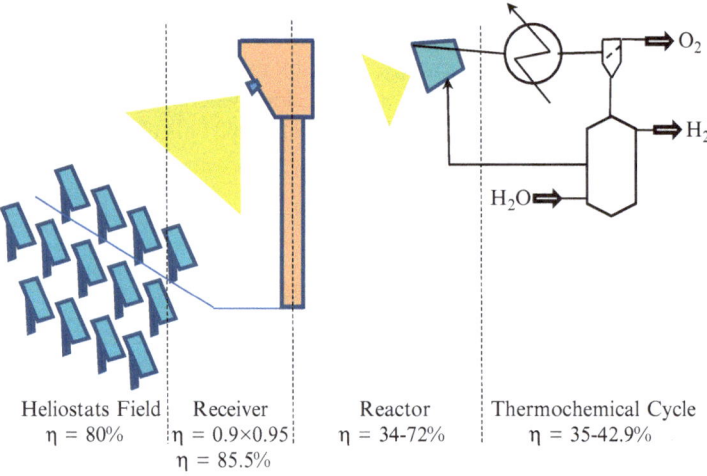

Fig. 5.2 Solar hydrogen production via thermochemical cycle (Modified from Charvin et al. [15])

The energy and exergy efficiency of the system can be calculated by considering the efficiencies of each component in the system. Various components, namely, heliostat field mirrors, receiver, reactor, and the thermochemical cycle are shown in Fig. 5.2. Here is a brief calculation for each component. To calculate the heliostat efficiency, the thermal energy available on the heliostat surface is an important parameter that can be calculated as

$$\dot{Q}_{Solar} = A_{mirrors} \times DNI \tag{5.1}$$

The usable part of this thermal energy, also called exergy, of the direct normal irradiance incident on the mirrors can be calculated as

$$\dot{Ex}_{Solar} = A_{mirrors}.DNI.\left(1 - \frac{T_a}{T_{Sun}}\right) \tag{5.2}$$

where $A_{mirrors}$ is the total area of heliostat mirrors (m^2) and DNI is direct normal irradiance (W/m^2).

This thermodynamic term resembles the Carnot cycle efficiency of a fictitious heat engine between sun and earth and limits the maximum usability of direct normal irradiance. The solar energy ($\dot{Q}_{Solar}$) captured by the heliostats is subjected to several losses before being available for chemical reaction; namely, the optical loss of the heliostat field from glass and atmospheric absorption, shadowing from nearby heliostats, and guidance errors ($\dot{Q}_{Heliostats}$); the optical loss at the receiver aperture, radiations split around the reactor aperture thus not absorbed by the receiver ($\dot{Q}_{Concentration}$); the optical loss from reflection or absorption by the glass window of the solar reactor ($\dot{Q}_{Window}$); the radiation emitted by the cavity ($\dot{Q}_{Reradiation}$); and conduction through the reactor walls ($\dot{Q}_{Wall}$).

The energy transferred to the reactants and available for the reaction ($\dot{Q}_{Chemical}$) is the difference between the energy captured by the heliostats ($\dot{Q}_{Solar}$) and the foregoing energy losses as follows [15]:

$$\dot{Q}_{Chemical} = \dot{Q}_{Solar} - \left(\dot{Q}_{Heliostats} + \dot{Q}_{Concentration} + \dot{Q}_{Window} + \dot{Q}_{Reradiation} + \dot{Q}_{Wall}\right) \tag{5.3}$$

A simple approach to calculate the energy available at the reactor for the chemical reaction is as follows. The thermal energy available at the receiver surface can be given as

$$\dot{Q}_{Rec} = F_{Rec}.A_{Apr}.DNI.C.\rho_{mirror} \tag{5.4}$$

where

ρ_{mirror} = Reflectivity of the mirror
A_{Apr} = Area of aperture (m^2)
C = Concentration factor
σ = Stefan–Boltzmann constant (5.67×10^{-8} W/m^{2}K^4)
F_{Rec} = Efficiency factor of receiver

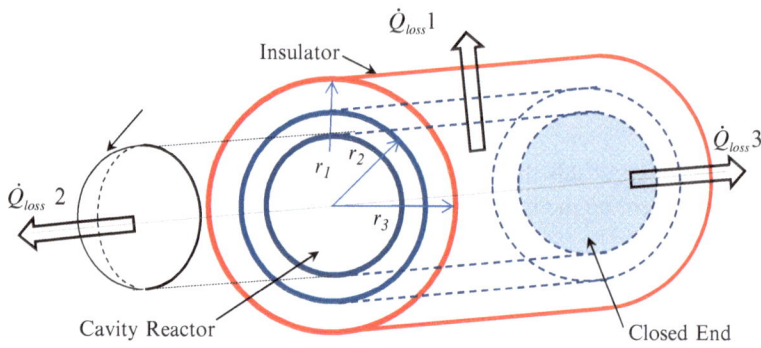

Fig. 5.3 Heat losses in cavity reactor

The thermal energy available at the receiver surface passes through a glass cover to enter the cavity reactor. As shown in Fig. 5.3, the reactor walls are properly insulated to minimize the heat loss but some losses are inevitable, namely, conductive and convective heat loss through reactor wall and insulation from the cylindrical surface, radiative heat loss through the aperture via the glass cover, and conductive and convective heat loss through the wall and insulation from the back surface of the reactor. Because the reactor/receiver temperature is very high, the radiative heat loss is dominant over the conductive and convective heat losses to ambient; therefore, the latter are neglected in the heat loss calculation through the aperture/glass cover. Therefore, the thermal energy available for the chemical reaction at the reactor can be given as

$$\dot{Q}_{Chemical} = \dot{Q}_{Rec}.\alpha.\tau - \dot{Q}_{loss}1 - \dot{Q}_{loss}2 - \dot{Q}_{loss}3 \tag{5.5}$$

$$\dot{Q}_{Chemical} = F_{Rec}.A_{Apr}.DNI.C.\rho.\alpha.\tau - U_C A_S.(T_{Rec} - T_a)$$
$$- \sigma \varepsilon A_{Apr}.(T_{Rec}^4 - T_a^4) - U_B.A_1.(T_{Rec} - T_a) \tag{5.6}$$

with

$$U_C = \cfrac{1}{\frac{A_1.\ln(r_2/r_1)}{2.\pi.k_1.L} + \frac{A_1.\ln(r_3/r_2)}{2.\pi.k_2.L} + \frac{A_2}{A_3}.\frac{1}{h_0}} \tag{5.7}$$

$$U_B = \cfrac{1}{\frac{(r_2-r_1)}{k_1} + \frac{(r_3-r_2)}{k_2} + \frac{1}{h_0}} \tag{5.8}$$

where

α = Absorptivity of reactor/cavity receiver
ε = Emissivity of the reactor/cavity receiver

$\rho =$ Reflectivity of mirrors

$\sigma =$ Stefan–Boltzmann constant (5.67×10^{-8} W/m^{2}K^4)

$\tau =$ Transmissivity of glass cover

$T_{\text{Rec}} =$ Absorber/receiver surface temperature (K)

$T_a =$ Ambient temperature (K)

$\dot{Q}_{loss}1 =$ Conductive and convective heat loss through cylindrical surface

$\dot{Q}_{loss}2 =$ Radiative heat loss through aperture

$\dot{Q}_{loss}3 =$ Conductive and convective heat loss through back surface at closed end

$U_C =$ Overall heat loss coefficient for cylindrical surface

$U_B =$ Overall heat loss coefficient for back surface at closed end

$k_1 =$ Thermal conductivity of reactor wall

$k_2 =$ Thermal conductivity of insulator

$A_S =$ Surface area of insulator

$r_1 =$ Inner radius of cavity reactor

$r_2 =$ Outer radius of cavity reactor

$r_3 =$ Outer radius of insulator

Here, absorptivity and emissivity of the reactor/cavity receiver are both assumed equal to unity [83, 150].

Generally, the energy lost at the reactor aperture, that is, $\dot{Q}_{Concentration}$, is fixed at 10 % of the energy reflected by the heliostat field, and the energy lost by reflection and absorption at the window ($\dot{Q}_{Window}$) amounts to 5 % of the radiation reaching the reactor aperture. In some cases, a window is not needed; therefore, $\dot{Q}_{Window}$ is not considered (e.g., for Fe$_2$O$_3$/Fe$_3$O$_4$ cycle) because the reduction reaction can be conducted under air without the use of any window. The efficiency of solar energy collection and concentration (sun to reactor efficiency) can be calculated by taking the product of the two efficiencies, namely, heliostat field efficiency (80 %) and receiver efficiency (90 % × 95 %). Therefore, the efficiency of solar energy collection and concentration (sun to reactor efficiency) is equal to 68.4 % (80 % × 90 % × 95 %). For a cavity reactor, the energy loss from radiation from the aperture of the reactor can be given as

$$\dot{Q}_{Reradiation} = \varepsilon . \sigma . A_{Apr} . \left(T_{Rec}^4 - T_a^4 \right) \tag{5.9}$$

Energy available at the aperture of the cavity reactor is the product of direct normal irradiance, concentration factor, and area of the aperture and can be given as

$$DNI.C.A_{Apr} = \dot{Q}_{Solar} - \left(\dot{Q}_{Heliostats} + \dot{Q}_{Concentration} + \dot{Q}_{Window} \right) \tag{5.10}$$

Therefore, the area of the aperture becomes

$$A_{Apr} = \frac{\dot{Q}_{Solar} - \left(\dot{Q}_{Heliostats} + \dot{Q}_{Concentration} + \dot{Q}_{Window} \right)}{DNI.C} \tag{5.11}$$

Substituting the area of the aperture in Eq. 5.3, energy lost from radiation becomes

$$\dot{Q}_{Reradiation} = \varepsilon.\sigma. \left[\frac{\dot{Q}_{Solar} - \left(\dot{Q}_{Heliostats} + \dot{Q}_{Concentration} + \dot{Q}_{Window} \right)}{DNI.C} \right] \cdot \left(T_{Rec}^4 - T_a^4 \right)$$

(5.12)

It can be inferred from Eq. 5.6 that the energy loss from radiation increases with the temperature of the reactor and decreases with the increase in concentration factor.

Three efficiencies that can be helpful to analyze the performance of the thermochemical-based solar hydrogen system are defined as follows [15]:

$$\eta_{Sun.to.Reactor} = 1 - \frac{\left(\dot{Q}_{Heliostats} + \dot{Q}_{Concentration} + \dot{Q}_{Window} \right)}{\dot{Q}_{Solar}}$$

(5.13)

$$\eta_{Reactor} = 1 - \frac{\left(\dot{Q}_{Reradiation} + \dot{Q}_{Wall} \right)}{\dot{Q}_{Solar} - \left(\dot{Q}_{Heliostats} + \dot{Q}_{Concentration} + \dot{Q}_{Window} \right)} = \frac{\dot{Q}_{Chemical}}{\dot{Q}_{Solar} - \left(\dot{Q}_{Heliostats} + \dot{Q}_{Concentration} + \dot{Q}_{Window} \right)}$$

(5.14)

$$\eta_{Cycle} = \frac{F_{H_2}.HHV_{H_2}}{\dot{Q}_{Chemical}}$$

(5.15)

where $\dot{Q}_{Chemical}$ is the energy transferred to the chemical reactants to run the thermochemical cycle. A fraction of this energy, equal to η_{Cycle}, is stored in hydrogen. The remaining energy, lost in the cycle ($\dot{Q}_{Cycle}$), is given by

$$\dot{Q}_{Cycle} = \dot{Q}_{Chemical}. \left(1 - \eta_{Cycle} \right)$$

(5.16)

The total solar power ($\dot{Q}_{Solar}$) collected by the heliostats is obtained from

$$\dot{Q}_{Solar} = \frac{F_{H_2}.HHV_{H_2}}{\eta_{Sun.to.Reactor} \cdot \eta_{Reactor} \cdot \eta_{Cycle}} = A_{mirrors}.DNI$$

(5.17)

The product of the three efficiencies just defined corresponds to the global efficiency (η_{Global}) of the hydrogen production system. This global efficiency can also be denoted as the energy efficiency of the system and can be given as

$$\eta = \frac{F_{H_2}.HHV_{H_2}}{\dot{Q}_{Solar}} = \frac{F_{H_2}.HHV_{H_2}}{A_{mirrors}.DNI}$$

(5.18)

Similarly, the exergy efficiency of the system can be calculated as

$$\psi = \frac{F_{H_2}.HHV_{H_2}}{\dot{Ex}_{Solar}} = \frac{F_{H_2}.HHV_{H_2}}{A_{mirrors}.DNI\left(1 - \frac{T_a}{T_{Sun}}\right)} \tag{5.19}$$

Considering the efficiencies of each component as shown in Fig. 5.2, the efficiency of the foregoing solar thermochemical hydrogen production system ranges between 8.14 % and 21.13 %.

5.3   Energy and Exergy Efficiencies of Solar H$_2$ Production Via Electrolysis Using Concentrated Solar Energy to Generate Electricity

The energy and exergy efficiency of the system can be calculated by considering the efficiencies of each component in the system. Here is a brief calculation for each component. To calculate the collector efficiency, the heat available on the collector surface is an important parameter that can be calculated by the Hottler Whiller equation as [151]

$$\dot{Q} = F_R A_A \left[\rho_R \alpha_A I_t C - U_L(T_A - T_a) - \varepsilon\sigma\left(T_A^4 - T_a^4\right)\right] \tag{5.20}$$

where

ρ_R = Reflectivity of the reflector surface
α_A = Absorptivity of the absorber surface
I_t = Total solar radiation (W/m^2)
A_A = Area of absorber surface (m^2)
C = Concentration factor
U_L = Heat transfer coefficient from absorber surface to ambient (W/m^{2}K)
T_A = Absorber/receiver surface temperature (°C)
T_a = Ambient temperature (°C)
ε = Emissivity of the absorber surface
σ = Stefan–Boltzmann constant (5.67 × 10^{-8} W/m^{2}K^4)
F_R = Efficiency factor

Useful heat available at the receiver surface is limited by the Carnot efficiency of a fictitious heat engine working between the temperature of the receiver and the ambient temperature and can be given as

$$\dot{Q}_u = \dot{Q}\left(1 - \frac{T_a}{T_A}\right) \tag{5.21}$$

The energy efficiency of the concentrating collector can be defined as the ratio of thermal energy available at the receiver surface to the incident solar energy on the concentrator and can be expressed as

$$\eta_C = \frac{\dot{Q}}{I_t A_C} \tag{5.22}$$

where A_C is the area of collector (m^2).

The exergy efficiency of the concentrating collector can be defined as the ratio of the product of useful thermal energy available at the receiver surface and the Carnot cycle efficiency to the incident solar exergy on the concentrator and can be expressed as

$$\psi_C = \frac{\dot{Q}_u}{I_t \left(1 - \frac{T_a}{T_S}\right) A_C} \tag{5.23}$$

The mechanical work (shaft work) of a heat engine can be calculated as

$$\dot{W} = \eta_{HE} \dot{Q}_u \tag{5.24}$$

where

$$\eta_{HE} = 1 - \frac{T_a}{T_A} \tag{5.25}$$

Note that the efficiency of the heat engine (η_{HE}) is the Carnot cycle efficiency that limits the maximum usage/application of thermal energy by any heat engine. In other words, no heat engine can perform better than its Carnot cycle efficiency. Therefore, the exergy efficiency of the heat engine is less than the Carnot cycle efficiency. Once mechanical work is produced, electricity can be generated by coupling an electrical generator/alternator with the heat engine. It can also be calculated by taking the ratio of electrical power generated to solar energy incident on the collector area. Hence, the energy efficiency of the electrical generator can be calculated as

$$\eta_{GEN} = \frac{P_E}{P_M} \tag{5.26}$$

where

P_E = electrical power (W), and
P_M = mechanical power (W).

The exergy efficiency of a electrical generator/alternator can also be considered to be the same as the energy efficiency because the electrical output and mechanical input remain the same. The efficiency of the electrical generator/alternator is limited by iron loss, copper loss, and the voltage drop in the diode bridge and is between 50 % and 62 % [152]. The final expression for exergy efficiency of electrical generator/alternator becomes

$$\Psi_{GEN} = \frac{P_E}{P_M} = \frac{VI}{\tau \times \omega} = \frac{VI}{\tau \times \frac{2\pi \times rpm}{60}} \qquad (5.27)$$

where

$\tau =$ Torque (Nm)
$\omega =$ Rotations per minute (rpm)
$V =$ Voltage (V)
$I =$ Current (A)

The energy efficiency of the electrolyzer can be calculated as

$$\eta_{EL} = \frac{\dot{m}_{H_2} HHV_{H_2}}{P_{in}} \qquad (5.28)$$

The exergy efficiency of the electrolyzer can be calculated as

$$\Psi_{EL} = \frac{\dot{Ex}_{out}}{P_{in}} = \frac{\dot{Ex}_{H_2} + \dot{Ex}_{O_2}}{P_{in}} \qquad (5.29)$$

When all the foregoing efficiencies of each unit are known, the overall energy efficiency of solar thermal hydrogen production system can be calculated as

$$\eta = \eta_C \times \eta_{HE} \times \eta_{GEN} \times \eta_{EL} \qquad (5.30)$$

Similarly, the overall exergy efficiency of the solar thermal hydrogen production system becomes

$$\Psi = \Psi_C \times \Psi_{HE} \times \Psi_{GEN} \times \Psi_{EL} \qquad (5.31)$$

5.4   Energy and Exergy Efficiencies of Solar H$_2$ Production Via Electrolysis Using the PV System to Produce Electricity

Similarly, the overall energy efficiency of a photovoltaic hydrogen production system can be calculated as

$$\eta = \eta_{PV} \times \eta_{CR} \times \eta_{IN} \times \eta_{EL} \qquad (5.32)$$

where

$$\eta_{PV} = \frac{V_{oc}I_{sc}}{I_t A} \qquad (5.33)$$

where $V_{oc} =$ open circuit voltage (V) and $I_{sc} =$ short circuit current (A).

And the overall exergy efficiency of photovoltaic hydrogen production system becomes

$$\psi = \psi_{PV} \times \psi_{CR} \times \psi_{IN} \times \psi_{EL} \tag{5.34}$$

where

$\psi_{PV} =$ Exergy efficiency of photovoltaic panel,
$\psi_{CR} =$ Exergy efficiency of charge regulator, and
$\psi_{IN} =$ Exergy efficiency of inverter.

The exergy efficiency of the photovoltaic panel can be given as

$$\psi_{PV} = \frac{FFV_{oc}I_{sc}}{I_t\left(1 - \frac{T_a}{T_S}\right)A} = \frac{VI}{I_t\left(1 - \frac{T_a}{T_S}\right)A} \tag{5.35}$$

where the fill factor (FF) is the ratio of electrical output to the maximum possible electrical output from a PV cell. Fill factor can be expressed as

$$FF = \frac{VI}{V_{oc}I_{sc}} \tag{5.36}$$

Chapter 6
Environmental Impact and Sustainability Assessment

6.1 Introduction

The world's energy demand is ever increasing, and to fulfill that demand we often look for engineering solutions, that is, conversion of energy from one form to another. We convert one form of energy to another convenient form to use the energy better. This statement can better be understood by the following example: we convert chemical energy of fossil fuels to heat (thermal energy) first, and then the thermal energy is converted to electrical energy for our convenience to use it in our homes, offices, industries, hospitals, schools, etc. Such energy conversion activities come under engineering activities. Engineering activities are often associated with some environmental problems/issues, mainly the injection of greenhouse gas into the environment. Environmental damage caused by the greenhouse effect further causes global warming, which is responsible for climate change. Therefore, it is important to know how adversely engineering activities are affecting our environment. Environmental impact assessment is simply an indicator of these concerns.

On the one hand, we have well-established technologies that use fossil fuels to produce electricity, which can further be utilized to convert hydrogen from water electrolysis, but in due course we damage our environment at a faster rate. One kilowatt-hour (kWh) of energy conversion from coal injects almost 0.98 kg CO_2 into the environment. On the other hand, renewable technologies are becoming popular as they cause less damage to environment, but they have their own limitations in terms of poor efficiencies, intermittency in availability of input energy, etc. Therefore, there is a strong need to understand and make a correct decision for the selection of engineering technology for energy conversion to protect our environment without compromising on our energy needs.

Sustainable development can be defined as a mode of human development in which we use our resources to meet the needs of present the generation without affecting the environment and the needs of future generation. Sustainability assessment is an indicator of sustainable development. Energy, environment, and

Fig. 6.1 The triangle of energy, environment, and sustainable development

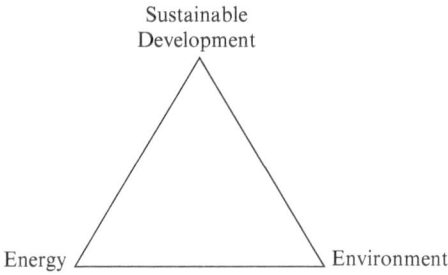

sustainable development can be referred to as three vertices of an equilateral triangle, as shown in Fig. 6.1. If you come close to any vertex, for example, energy conversion, you move farther away from the other two points (i.e., environment and sustainability). Therefore, it has become very important for energy engineers and environmentalists to find the proper balance among the three factors.

6.2   Environmental Impact Reduction Factor

As already mentioned, electricity is one of the energy inputs for hydrogen production; therefore, its generation cost, efficiency, and environmental impact become important parameters for environmentally benign and cost-effective hydrogen production. Table 6.1 shows these parameters along with the efficiency of a hydrogen production system using an electrolyzer unit and the environmental impact reduction factor.

The efficiency of hydrogen production is calculated by multiplying the efficiencies of electricity production and the electrolyzer unit (i.e., 52 %) [10], whereas the environmental impact reduction factor (*EIRF*) can be given as

$$EIRF = \frac{gCO_{2coal} - gCO_2}{gCO_{2coal}} \tag{6.1}$$

Here, the numerator denotes the difference in carbon dioxide injection into the environment by conventional (coal-based) and nonconventional methods, and the denominator denotes carbon dioxide injection into the environment by coal-based electricity generation. No carbon dioxide is produced during the electrolysis of water, but it is produced during electricity production. Hence, *EIRF* is calculated by using the carbon dioxide emission by different sources for electricity production only. The value of *EIRF* remains between 0 and 1, where 1 represents the best technology and 0 the worst technology, for which the environmental impact is highest and lowest, respectively. Table 6.1 shows that conventional technologies such as coal and gas are more economical for per kWh electricity generation but they pollute the environment more. However, the price of per kWh electricity generation for coal and gas is lower than for renewable energy sources. Therefore,

Table 6.1 Mean price of electricity generation, efficiencies of electricity and hydrogen generation, average greenhouse gas emissions expressed as CO_2 equivalent, and environmental impact reduction factor (EIRF) for individual energy-generation technologies

| Energy sources | US$/kWh | Efficiency | | gCO₂/kWh | EIRF |
		Electricity (%)	Hydrogen production[a] (%)		
Photovoltaic	0.24	4–22 %	2–12 %	90	0.91
Wind	0.07	24–54 %	13–28 %	25	0.98
Hydro	0.05	>90 %	47 %	41	0.96
Geothermal	0.07	10–20 %	5–11 %	170	0.83
Coal	0.042	32–45 %	17–23 %	1,004	0.00
Gas	0.048	45–53 %	23–28 %	543	0.46

Source: Modified from Evans et al. [153]
[a]Hydrogen production considered here is by using electrolyzer unit only

Table 6.2 Sustainability indicators for some renewable energy sources

	Photovoltaic	Wind	Hydro	Geothermal
Price	4	3	1	2
CO₂ emission	3	1	2	4
Availability and limitations	4	2	1	3
Efficiency	4	2	1	3
Land use	1	3	4	2
Water consumption	2	1	3	4
Social impact	2	1	4	3
Total	20	13	16	21

Source: Evans et al. [153]

one can say that renewable energy sources are either cost effective (for example, wind, biomass, geothermal) or less polluting to the environment. Further, electricity generation efficiency of the different technologies, also given in Table 6.1, shows clearly that some renewable sources (for example, wind and hydro power), if not possessing better efficiency, are competitive with the conventional technology (coal, gas, etc.). The efficiency of hydrogen production is greatly affected by the efficiency of electricity production. One can see from Table 6.1 that by using photovoltaic, wind, and geothermal energy, the hydrogen production efficiency can be 2–12 %, 13–28 %, and 5–11 %, whereas for hydro it is highest, i.e., 47 %, and for coal and gas it remains between 17–23 % and 23–28 %, respectively. However, for the environmental impact reduction factor, coal is the most polluting (0.00 as coal is taken as base case) and gas is second highest (0.46). Wind, hydro, photovoltaic, and geothermal energy are said to be more benign environmentally and are sustainable sources, as the environmental impact reduction factor for each is quite high, 0.98, 0.96, 0.91, and 0.83, respectively.

Evans et al. [153] have given sustainability indicators for some renewable technologies, namely, wind, hydro, photovoltaic, and geothermal: each technology was ranked from 1 to 4 according to the corresponding indicator (as shown in Table 6.2), with 1 being the best technology for that indicator. The average and range were considered together, where values were quantifiable, as there was often significant overlap between values. Some impact categories, such as availability and limitations as well as social impacts that cannot be quantified, were assessed qualitatively. In case of limitations, hydro was chosen as the least limited, because of its ability to provide base load power, number of suitable sites worldwide, and flexibility of operation. Wind was considered the second best for similar reasons. Geothermal is slightly more limited worldwide, with fewer suitable locations. Solar is considered the most limited, because excess power acquired during daylight hours cannot yet be stored efficiently to provide adequate power during off-sunshine periods (nights and on cloudy days). As far as social impacts were concerned, wind was allocated the least negative social impacts, because of its benign nature. Solar was second, as careful management during manufacture and proper site selection mitigate its potential negative impacts, and geothermal was third consequent to increased seismic activity and pollution potential. Hydro had the largest impact, primarily the result of the large numbers of people and animals displaced during inundation following dam development. The ranking in Table 6.2 suggests electricity production from wind is the most sustainable, followed by hydropower, and then solar and geothermal were found to rank the lowest from the four non-combustion renewable energy technologies [153]. This ranking was provided for global international conditions, although each technology can be significantly geographically affected. For a certain geographic location, some of the listed sustainability indicators may become more important than others.

6.3   Sustainability Index

Sustainable development requires not only that sustainable energy resources be used, but also the resources should be used efficiently. Exergy analysis is essential to improve efficiency, which allows society to maximize the benefits it derives from its resources while minimizing the negative impacts (such as environmental damage). Exergy efficiency not only gives an idea of the amount (quantity) of useful energy that can be completely utilized in useful work but also the quality of energy. The part of energy that is not going to be utilized for useful work is called exergy destruction. Exergy analysis gives a realistic analysis of a system for its possible feasibility and the exergy destruction gives the scope for improvement in the existing system. Another way to understand the scope of improvement or the performance of a system is through the sustainability index that tells how sustainable a system is in actual practice.

Once the exergy efficiency of the system is known, the sustainability index can be calculated. The relation between exergy efficiency (ψ) and the sustainability index (SI) can be given as [154]:

$$\psi = 1 - \frac{1}{SI} \tag{6.2}$$

where

$$SI = \frac{1}{D_P} \tag{6.3}$$

and D_P is the depletion factor/number defined by Connelly and Koshland [155] as the ratio of exergy destruction rate to the input exergy rate to the system and can be given as

$$D_P = \frac{\dot{E}x_D}{\dot{E}x_{in}} \tag{6.4}$$

6.4 Sustainable and Environmentally Benign Hydrogen Production by Artificial Photosynthesis

Sustainable development requires not only that the sustainable energy resources be used, but also the resources should be used efficiently. Artificial photosynthesis, a concept presented by some authors (e.g., Collings and Critchley [156]), not only produces hydrogen (and food) but also is beneficial to the environment by removing carbon dioxide and adding oxygen. The concept is a replica of the photosynthesis of plants and algae. The two reactions involved in photosynthesis, as described earlier, are the light reaction and the dark reaction. To understand the concept better, let us understand these processes and their end products.

(a) Light reaction

In this reaction, light energy is absorbed by special cell membranes and transferred to chlorophylls. Here electrochemical reactions commence that generate vital "energy-rich" biological compounds. Oxygen is produced as a by-product in this process and is released to the atmosphere. This process is actually nature's own photovoltaic energy conversion system (photo-systems), in which the trapped light energy is first converted into electrically stored energy in cell membranes. The light phase requires the cooperation of membrane-bound photochemical assemblies (also called photo-systems). Each photo-system operates in series, to photochemically "charge" the membrane. The anodic reaction for the electrolysis can be expressed as

$$2H_2O = 4H^+ + O_2 + 4e^- \tag{6.5}$$

(b) Dark reaction

In the dark reaction, the products of the light phase, that is, energy-rich biological compounds, are used within cells for the formation of carbohydrate (sugars) from carbon dioxide, via a series of biochemical intermediates in the presence of some enzymes (catalysts) This process is central to the progressive chemical "assembly" of sugar molecules from carbon dioxide and water. The latter is the only point at

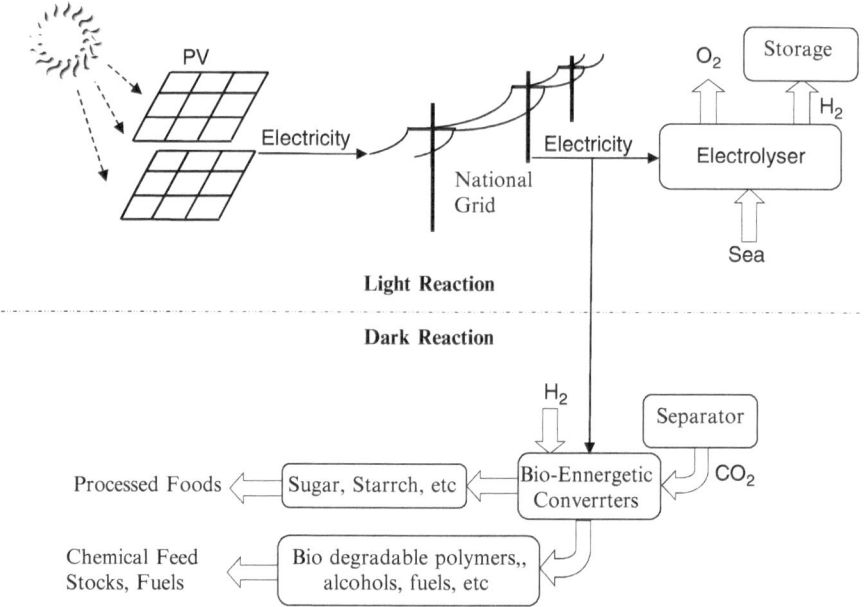

Fig. 6.2 Schematic of artificial photosynthesis concept (Modified from Collings and Critchley [156])

which continued water consumption is absolutely required for carbohydrate formation and represents only a tiny fraction of the water applied conventionally in plant growth.

So, the overall reaction of the natural photosynthesis including the carbon fixation process can be given as

$$CO_2 + H_2O + Sun\ Light = O_2 + Carbohydrate \qquad (6.6)$$

Artificial photosynthesis is a mean of achieving these aims of clean power generation and dry food production as stated earlier. Figure 6.2 shows a schematic diagram of artificial photosynthesis. The figure is separated by a dashed line: the upper half shows the light reaction and the bottom half the dark reaction of photosynthesis. The main steps in the natural photosynthesis processes of plants and bacteria provide the models and inspiration for a totally biomimetic, industrial-scale technological approach to achieve the following specific goals [156].

- Electricity generation using photovoltaic systems. The power generated can directly be supplied to the national grid.
- "Dry agriculture," employing enzyme bed reactor systems to fix carbon dioxide from the air or other convenient sources, powered by hydrogen and bioelectric transducers drawing power from the national grid. These systems will produce carbohydrates (food), liquid fuels, chemical feed

stocks, and polymers for fiber production. Water usage will be at or near the absolute chemical minimum and thousands of times lower than in conventional agriculture.

- Hydrogen production from seawater or other suitable water sources. Electrode systems employing catalytic surfaces modeled on the relevant high-efficiency active sites in photosynthetic organisms will achieve the electrolytic decomposition of water (into hydrogen and oxygen).

Chapter 7
Case Studies

7.1 Introduction

In this section, some case studies are presented to analyze the performance of some solar hydrogen production systems on the basis of their exergy and energy efficiencies, sustainability, and environmental impact. The section is subdivided into three parts: first, exergy and energy analysis; second, the sustainability of different systems; and third, the environmental impact of a natural gas-based hydrogen production system.

7.2 Case Study 1

In this case study, the analysis presented in the Chapter 5 is applied to photovoltaic and solar thermal hydrogen production systems. Case study 1 is divided into three parts. The first part is an efficiency analysis of the concentrating collector for the solar thermal hydrogen production system. The second part concerns the overall energy and exergy efficiency analysis of thermal and PV hydrogen production systems, and the third part concerns the sustainability index analysis of these two systems. To analyze the solar thermal hydrogen production system, the energy and exergy efficiency analysis of concentrating collector is very important, because this component is responsible for converting solar energy to concentrated thermal energy; it is affected by climatic parameters such as intensity of solar radiation and ambient temperature and physical parameters of the concentrator such as reflectivity of reflector surface, absorptivity, and emissivity of absorber surface, areas of absorber surface and collector, concentration ratio, and overall heat loss coefficient. For the energy and exergy analysis of concentrating collector of solar thermal system, the data of Tyagi et al. [151] are adopted and presented in Table 7.1.

I. Dincer and A.S. Joshi, *Solar Based Hydrogen Production Systems*,
SpringerBriefs in Energy, DOI 10.1007/978-1-4614-7431-9_7, © The Author(s) 2013

Table 7.1 Energy and exergy efficiency
parameters for solar concentrating
collector

Parameter	Value
ρ_R	0.9
α_A	0.9
A_A	0.8 m^2
C	100
F_R	0.9
ε	0.9
A_C	80 m^2
U_L	8 W/m^2 K

Source: Tyagi et al. [151]

Table 7.2 Modified parameters for the
evaluation of energy and exergy efficiency
of solar concentrating collector

Parameter	Value
ρ_R	0.9
α_A	0.8
A_A	0.0079 m^2
C	250
F_R	0.9
ε	0.2
A_C	1.96 m^2
U_L	25 W/m^2 K

Source: Tyagi et al. [151]

It should be noted that the absorptivity and emissivity of the absorber surface are assumed to be each equal to 0.9. To present a more realistic analysis and for comparison purposes, some of the data from Table 7.1 such as absorptivity, emissivity, areas of absorber surface and collector, concentration ratio, and overall heat loss coefficient have been modified and are given in Table 7.2. To analyze solar concentrator performance on the basis of energy and exergy efficiency, four different cases are described in brief as follows:

Case 1: Uses the data given in Table 7.1.
Case 2: Uses the same data as Case 1 except for the value of emissivity.
Case 3: Uses the same data as Case 1 except for the values of absorptivity and
      emissivity.
Case 4: Uses the modified data as given in Table 7.2.

For analysis of the photovoltaic (PV) hydrogen system, the experimental setup of Yilanci et al. [10] has been adopted. A brief description of the PV hydrogen production system is as follows. The system was installed in 2007 in the Clean Energy Center (CEC) of Pamukkale University in Denizli, Turkey. The system is equipped with 5 kWe PV panels. Its simplified diagram is shown in Fig. 7.1. For performance investigation and assessment, one-half of the photovoltaic modules are on fixed tilt, and the other half are mounted on solar trackers. The fixed tilt (45° south) photovoltaic modules are located on the roof of the building. Each tracker consists of ten modules with the nominal power of 1.25 kWe.

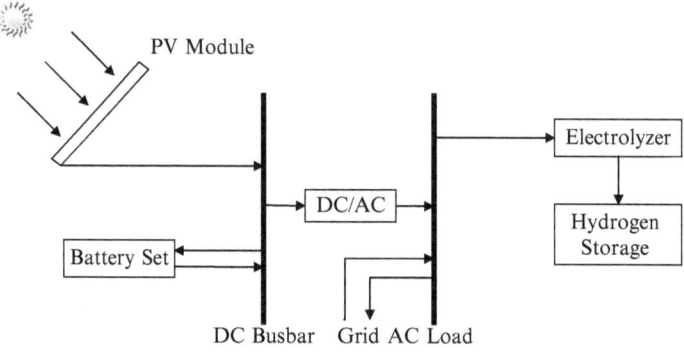

Fig. 7.1 Photovoltaic (PV)-driven water electrolysis system (Modified from Yilanci et al. [10])

Table 7.3 Energy and exergy efficiencies of photovoltaic (PV) hydrogen system components

Components	Energy efficiency (%)	Exergy efficiency (%)
PVs	11.2–12.4	9.8–11.5
Charge regulators	85–90	85–90
Inverter	85–90	85–90
Electrolyzer	56	52

Source: Modified from Yilanci et al. [10]

A line feed deionizer was selected to supply the quality of water needed for the electrolyzer. A basic particle water filter was also used before the deionizer. A PEM-type electrolyzer was used in the system. A metal hydride (MH) storage tank has been used for hydrogen. The downside is that the hydrogen produced for MH storage must be of very high purity. Six OVONIC 85G250B storage tanks were chosen. The system also includes a fuel cell to generate electricity from the produced hydrogen and is connected to the grid. Table 7.3 lists the efficiencies of some of the individual system components that are used to produce hydrogen. Average hydrogen production of the system for a week is 4.43 kg. The data presented in Table 7.3 would be used for energy and exergy analysis of the system.

The foregoing four cases are discussed with their explanation in detail and the different climatic parameters in the next subsection. The analyses presented in the previous sections and the foregoing data are applied to solar thermal and photovoltaic hydrogen production systems, and the results and their pertinent discussion for all three parts as described in the beginning of the case studies are presented as follows.

– Efficiency Analysis for Concentrating Collector. For the solar thermal hydrogen production system, first the energy and exergy efficiencies of the solar collector are evaluated by using Eqs. 5.20–5.23 and the data presented in Table 7.1. The effect of variation in solar radiation on the energy and exergy efficiency of concentrated collector is also calculated and is shown in Table 7.4 and Fig. 7.1,

Table 7.4 Effect of solar radiation, absorptivity and emissivity on the energy and exergy efficiency of a solar concentrating collector

I_t, W/m^2	Case 1 $\alpha_A = 0.9, \varepsilon = 0.9$		Case 2 $\alpha_A = 0.9, \varepsilon = 0.1$		Case 3 $\alpha_A = 0.8, \varepsilon = 0.2$		Case 4[a] $\alpha_A = 0.8, \varepsilon = 0.2$	
	η_C (%)	ψ_C (%)	η_C (%)	ψ_C (%)	η_C (%)	ψ_C (%)	η_C (%)	ψ_C (%)
500	23.9	16.07	60.74	40.79	48.04	32.26	51.76	34.76
750	40.25	27.03	64.79	43.51	53.62	36.01	56.14	37.7
1,000	48.41	32.51	67	44.87	56.42	37.89	58.33	39.18

[a]For Case 4, complete data are shown in Table 7.4

respectively. The effect of absorptivity and emissivity of the absorber surface on energy and exergy efficiency is also shown in Table 7.4. Table 7.4 shows four cases for energy and exergy efficiency of the solar concentrating collector, described as follows: Case 1 uses the data of Tyagi et al. [151], and Cases 2 and 3 use modified data from Tyagi et al. [151]. It should be noted that Tyagi et al. [151] have assumed the absorptivity and emissivity of the absorber surface each equal to 0.9 (Case 1).

To minimize the radiation losses, the emissivity is reduced in Cases 2, 3, and 4. In Case 3, the absorptivity and emissivity are taken as 0.8 and 0.2, respectively. Case 4 shows the effect of concentration factor, areas of absorber and collector surface, and overall heat loss coefficient on the efficiencies of the collector. The modified data for Case 4 are shown in Table 7.2. It should be noted that the absorber area is reduced to reduce the heat loss from radiation and the concentration factor is increased to have more thermal energy available at absorber surface. A more realistic value of overall heat loss coefficient (25 W/m^2 K) is considered at higher absorber temperatures; for example, at 823 K, the heat transfer coefficient is also high [157]. For calculations, three different solar intensities, that is, 500, 750, and 1,000 W/m^2, are considered, and the ambient and absorber temperatures are assumed as 298 and 823 K, respectively.

Figure 7.2 shows the variation of energy and exergy efficiency of a solar concentrating collector. It is clear from this figure that energy efficiency is higher than the exergy efficiency in all four cases, because the exergy efficiency, also called second law efficiency, incorporates the losses from irreversibility and hence gives a more reliable analysis. Energy and exergy efficiency increases with the increase in solar radiation because higher solar radiation is responsible for high thermal energy available at the absorber surface. Therefore, it can be said that the higher the intensity of solar radiation, the higher would be the energy and exergy efficiency of the concentrating collector. Further, comparing Case 1 and 2 it can be said that the energy and exergy efficiencies of the former are lower because of higher emissivity. Comparing Cases 2 and 3, it can be said that the efficiencies of Case 2 are higher because of low emissivity and high absorptivity. Comparing Cases 3 and 4, one may see that the efficiencies of Case 4 are higher because of a higher concentration factor. The area of absorber surface is reduced to reduce the heat losses caused by radiation.

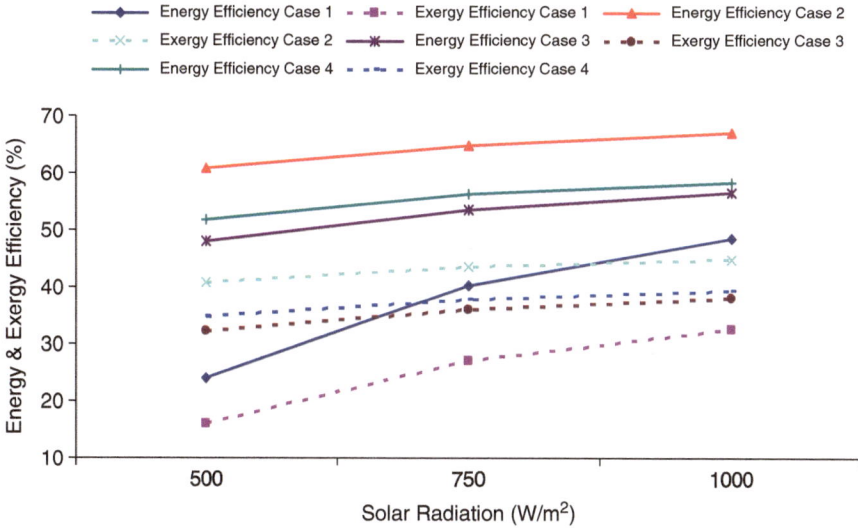

Fig. 7.2 Variation of energy and exergy efficiency with solar radiation for Cases 1 through 4 (Modified from Joshi et al. [158])

Table 7.5 Energy and exergy efficiencies of various components of a solar thermal hydrogen production system

Components	Energy efficiency (%)	Exergy efficiency (%)
Solar collector	58.33	39.18
Heat engine	63.8	30–50
Generator	50–62	50–62
Electrolyzer	56	52

Sources: Yilanci et al. [10]; Crouse and Anglin [152]

– Overall Efficiency Analysis for Solar Thermal and PV Hydrogen Production Systems. When the solar concentrating collector efficiency has been evaluated, the overall energy and exergy efficiencies of the solar thermal hydrogen production system can be calculated by using Eqs. 5.22, 5.23, 5.24, 5.25, 5.26, 5.27, 5.28, and 5.29. However, for the present case study, the energy and exergy efficiency data for other components (except for the storage system) (Table 7.5) of the solar thermal hydrogen production system as shown in Fig. 4.25 are considered. It is found that the overall energy and exergy efficiency of the solar thermal hydrogen production system ranges between 10.42–12.92 % and 3.06–6.31 %, respectively (Fig. 7.3) [158]. The minimum overall energy and exergy efficiencies are calculated by taking least values of individual components, and maximum value is calculated considering the maximum values of the same. Note that the energy efficiency of the heat engine is calculated as 63.8 % by using Eq. 5.25, which is also the Carnot cycle efficiency working between the higher (823 K) and lower (298 K) temperatures, and

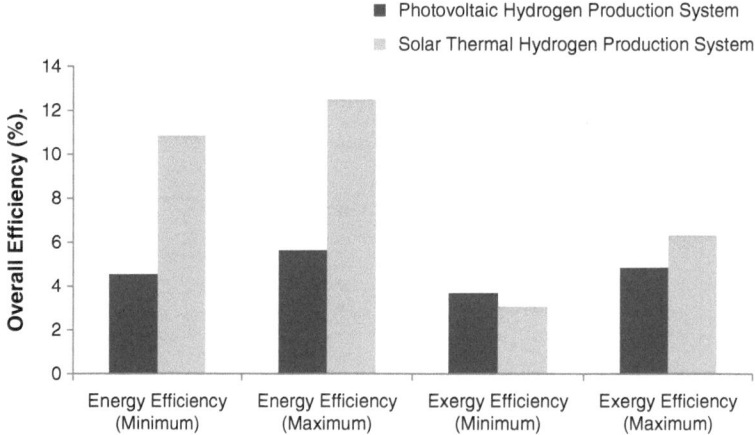

Fig. 7.3 Overall energy and exergy efficiency range for photovoltaic and solar thermal hydrogen production systems (Modified from Joshi et al. [158])

exergy efficiency of the heat engine is assumed as ranging between 30 % and 50 %, which is less than that of the Carnot cycle efficiency (also given in Table 7.5).

The energy and exergy efficiencies of the solar concentrating collector are taken from Case 4, corresponding to the solar radiation of 1,000 W/m². It is to be noted that these calculations are based on theoretical data and therefore give a theoretical analysis of energy and exergy efficiencies of a solar thermal hydrogen production system.

The overall efficiencies of a photovoltaic hydrogen production system are also shown in Fig. 7.3. The exergy and energy efficiencies of different components as given in Table 7.3 are considered to evaluate the overall exergy and energy efficiencies of the photovoltaic hydrogen production system by using Eqs. 5.32, 5.33, 5.34, 5.35, and 5.36. Referring to Table 7.3 and by using Eq. 5.32, the energy efficiency of the photovoltaic hydrogen production system is calculated as ranging between 4.53 % and 5.62 % (Fig. 7.3).

Similarly, the exergy efficiency of the photovoltaic hydrogen production system is calculated by using Eq. 5.34 as ranging between 3.68 % and 4.84 % (Fig. 7.3). It can be seen from Tables 7.3 and 7.5 that the energy and exergy efficiencies of individual components of the systems are higher as compared to the overall energy and exergy efficiencies of the system because the latter is the product of former and some components, namely, the solar concentrating collector and photovoltaic panels, have lower efficiencies.

An interesting point worth mentioning in Fig. 7.3 is that the minimum exergy efficiency of the solar thermal hydrogen production system is lower than the PV hydrogen production system. This difference may be because at low operating conditions, for example, in mornings, the intensity of solar irradiance is low and the ambient temperature is also low: this affects adversely the exergy efficiency of the former as the heat engine's efficiency drops because insufficient thermal energy is available with the low thermal gradient. However, these conditions are favorable for the PV system to maintain its electrical efficiency for the same solar irradiance. On the other hand, when the operating conditions are high, for example, at noontime, the intensity of solar irradiance and the operating temperature both are higher, therefore producing a higher thermal energy/gradient to the heat engine; obviously, the heat engine thermal efficiency is now better and hence the performance of the solar thermal hydrogen production system is higher at the maximum exergy efficiency level as compared to the PV hydrogen production system, which suffers lower electrical efficiency because of high operating temperatures.

7.3 Case Study 2: Sustainability Analysis of Solar Thermal and PV Hydrogen Production Systems

The sustainability index for the solar thermal hydrogen production system is calculated based on the overall exergy efficiency of the solar thermal hydrogen production system and by using Eqs. 6.2, 6.3, and 6.4 (shown in Fig. 7.4). It is clear from the same figure that sustainability index ranges between 1.03 and 1.07. A curve based on the theoretical calculation of the sustainability index is also shown to provide a clear understanding of its variation with exergy efficiency. The relationship between sustainability and exergy efficiency is given by Eq. 6.2. Here sustainability is measured in terms of sustainability index (SI).

It is clear from Fig. 7.4 that the sustainability index increases with increasing exergy efficiency. A system is said to be more sustainable if its sustainability index is higher, which is possible only when its exergy efficiency is also higher. The

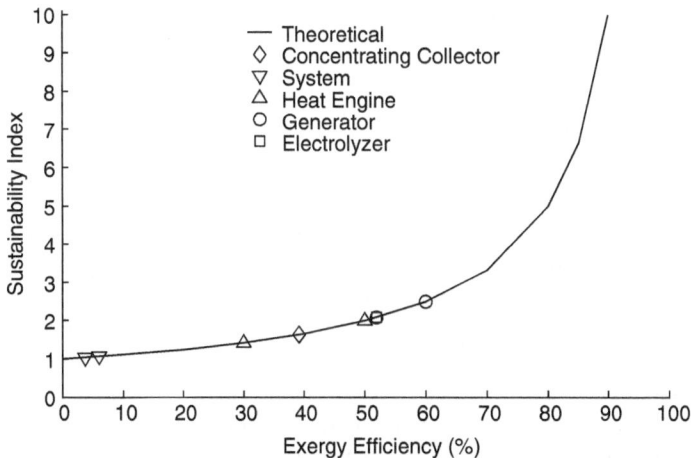

Fig. 7.4 Variation of sustainability index with exergy efficiency of a solar thermal hydrogen system (Modified from Joshi et al. [158])

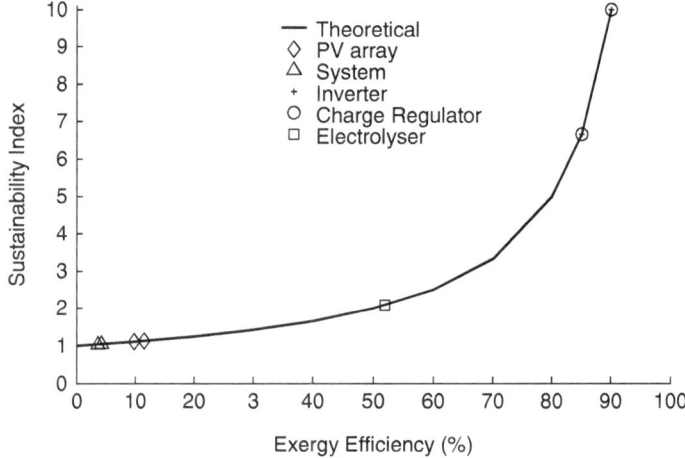

Fig. 7.5 Variation of sustainability index with exergy efficiency of a solar PV hydrogen system (Modified from Joshi et al. [158])

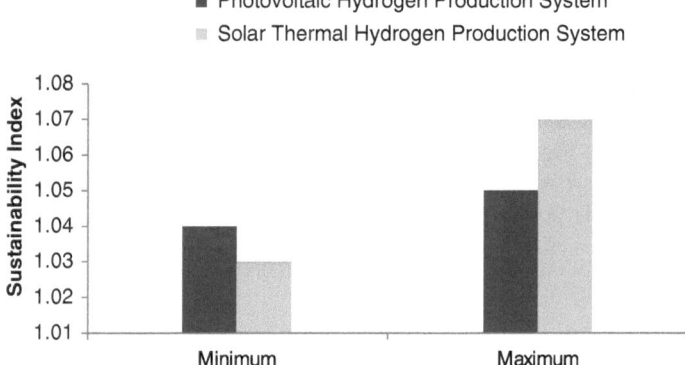

Fig. 7.6 Comparison of sustainability index for solar thermal and photovoltaic hydrogen production systems (Modified from Joshi et al. [158])

sustainability index of each component is also shown in Fig. 7.4. It can be easily understood that although the sustainability index of each component is higher individually, the sustainability index of the overall system is lower, because the exergy efficiency of the overall system is lower as it is the product of the exergy efficiencies of the different components of the system. The low sustainability index of the system is a consequence of its low overall exergy efficiency.

The sustainability index corresponding to the exergy efficiency of the photovoltaic hydrogen production system is calculated as ranging between 1.04 and 1.05 (Fig. 7.5). A lower value of sustainability index can be seen in Fig. 7.5, the result of the lower value of exergy efficiency of the system, which in turn is a result of the low exergy efficiency of the PV unit. It is clear from Figs. 7.4 and 7.5 that to improve the sustainability index one has to improve the exergy efficiency of the system, which can be done by system modifications and proper implementation and use of a particular technology.

A comparison of the sustainability index of both systems is shown in Fig. 7.6. It is clear from this figure that the sustainability index of a solar thermal hydrogen system is higher as compared to the photovoltaic hydrogen production system as the former has higher exergy efficiency.

7.4   Case Study 3: Environmental Impact of Methane-Based Hydrogen Production Systems

In this case study two types of hydrogen production systems have been considered, based on their input energy sources for their endothermic chemical reaction, for example, (i) natural gas based and (ii) renewable energy based. When natural gas (NG) is used as the input source, it represents both the reactant and the external power source (for steam generation, reforming furnace, purification, and auxiliary units); the actual consumption of natural gas depends on the energetic efficiency of the process. Figure 7.7 shows the natural gas consumption for a large-scale plant with typical 70–85 % energetic efficiency.

Methane consumption and the resulting CO_2 release are in the range of 3.0–3.7 kg CH_4/kg H_2 and 8.3–10.1 kg CO_2/kg H_2, respectively [73]. If an external non-fossil fuel and carbon-free power source is used instead of natural gas to supply process heat, for example, solar energy, methane consumption and CO_2 emissions

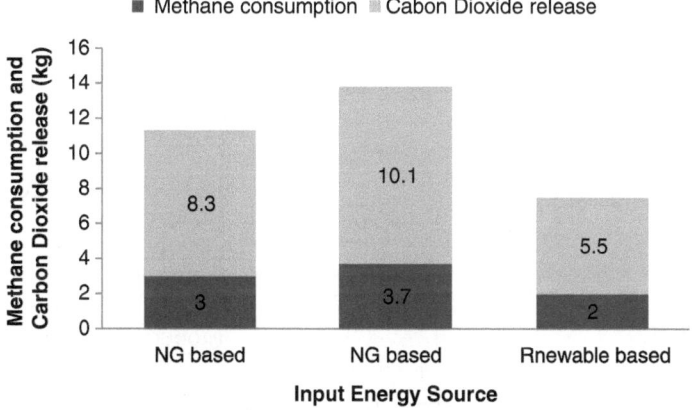

Fig. 7.7 Methane consumption and carbon dioxide release per kilogram (kg) of hydrogen production in a steam methane reforming (SMR) plant with different input energy sources. *NG*, natural gas (Modified from Giaconia et al. [73])

will result exclusively from reaction stoichiometry only, with specific values of 2.0 kg CH_4/kg H_2 and 5.5 kg CO_2/kg H_2, respectively [73]. It is clear from Fig. 7.7 that switching to renewables from fossil fuels reduces greenhouse gas (CO_2) emission into the environment by almost 50 %. There is also a reduction of natural gas consumption of 33 % to 54 %. Therefore, by using renewable energy sources, not only can the environmental impact can be reduced by a considerable reduction in greenhouse gas (GHG) release, but also the fossil fuels (NG in this case) can be conserved for future use. Furthermore, the improvements in the energetic/exergetic efficiency of the reaction may also decrease the methane consumption and result in further reduction in environmental impact.

7.5   Case Study 4: Energy Recovery Routes for Electro-Thermal Hydrogen Production

Even though the energy conversion for its use is of prime importance, energy recovery from waste for its reuse is also equally important. Energy recovery can be done for both high-grade and low-grade temperature heat. Some examples of high-grade temperature heat are municipal waste incineration and landfill gas combustion. An example for low-grade temperature heat includes heat released by various industries to the surroundings. Generally, such heat comes at lower grade, typically from 60 °C to 150 °C, or intermediate temperatures (up to 400 °C).

Figure 7.8 shows some possible routes to generate hydrogen through hybrid processes starting from energy recovery. The recovered energy is first converted to

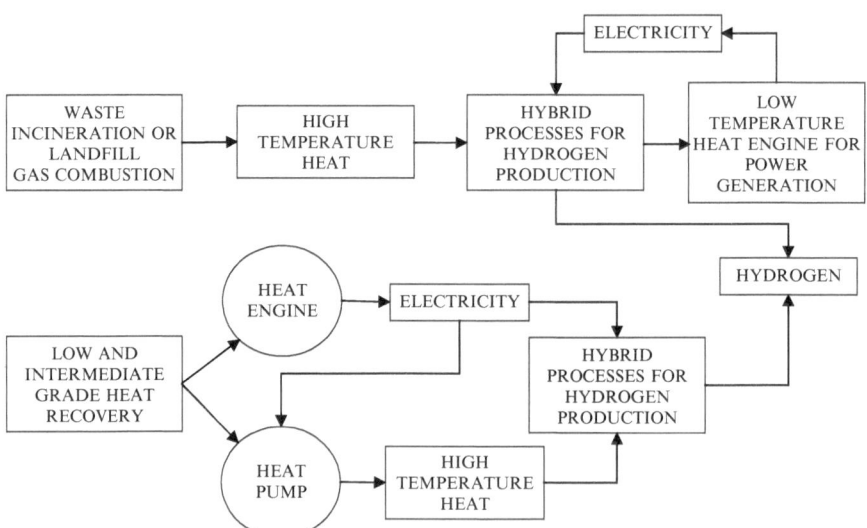

Fig. 7.8 Routes to hybrid hydrogen production processes driven by energy recovery (Modified from Dincer [19])

either high-grade or low/intermediate-grade energy. Further, the high-grade heat can be utilized either directly to produce hydrogen, or it can be used first to produce electricity by heat engines, and then the electricity can be utilized to produce hydrogen via some suitable hybrid process. The low/medium-grade heat can be used to produce electricity via heat engines and then the electricity can further be utilized to produce hydrogen, as already mentioned. Further, heat pumps can also be used to convert low/medium-grade heat to high-grade heat, and high-grade heat can be utilized to produce hydrogen, as discussed earlier. A combination of both electricity generation and converting low/medium-grade heat to high-grade heat may be done simultaneously to have economical hydrogen production via some suitable hybrid process. The economic advantage of the system mainly depends on the temperature level of the heat source. For example, if the level of temperature of heat source is too low, upgrading the temperature is no more beneficial [19].

7.6   Case Study 5: Life-Cycle Assessment of Solar-Assisted Hydrogen Production Systems

7.6.1   Introduction

Many environmental issues such as global warming and acidification are related to the production, transformation, and utilization of fossil fuels [159]. The risk of global climate change is of great concern to policy makers and the public. The relationship between the energy generation sector and environmental impact is being carefully considered in both industrialized and nonindustrialized countries [160]. Environmentally benign technologies are being developed to help ensure that future generations have cleaner energy systems and a more sustainable economy. The energy carrier hydrogen can facilitate improved environmental performance and sustainability of energy systems [19, 161]. Although addressing future energy challenges requires numerous measures and approaches, hydrogen is expected by many to play a major role, in part because it does not emit greenhouse gases (GHGs) during oxidation.

The energy carrier hydrogen is expected by many to become an important fuel that will help in solving several energy challenges we face today because its oxidation does not emit GHGs and does not contribute to climate change, provided it is derived from clean energy sources. Numerous researchers anticipate that hydrogen will replace petroleum products for fueling of transportation vehicles, in turn decreasing the dependence on petroleum. Industrial sectors, such as petro-chemical, agricultural, food processing, plastics, and manufacturing, use hydrogen heavily as a commodity. Hydrogen complements the energy carrier electricity, which can be generated from a variety of primary energy sources and is widely used in a broad range of applications. These two energy carriers are expected to have complementary roles in the future, in part because hydrogen adds the

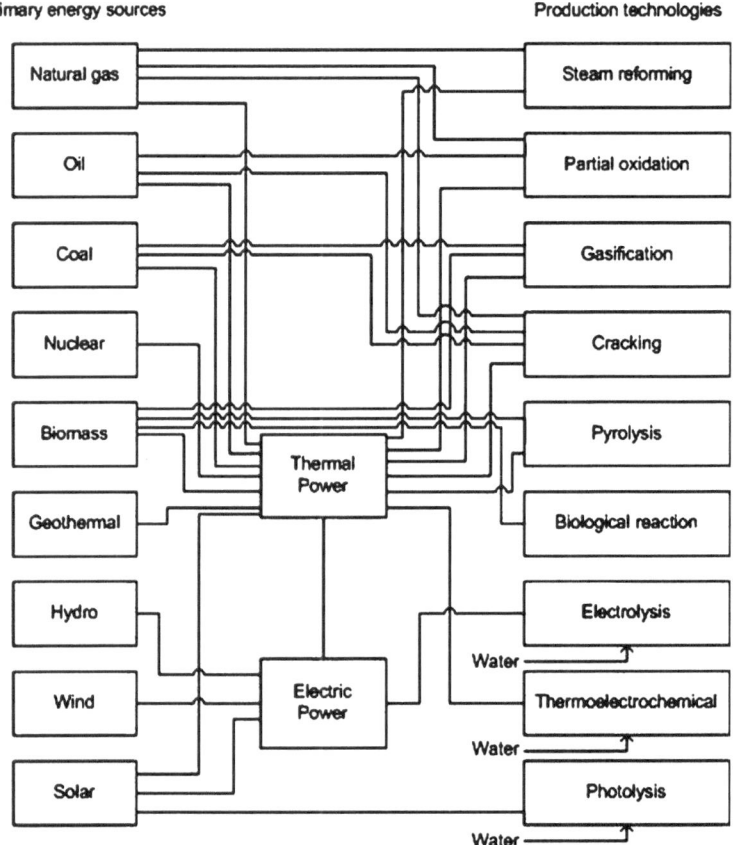

Fig. 7.9 Principal alternative methods of hydrogen production from various energy sources (Modified from Boehm et al. [163])

capability of storage. Hydrogen storage is a promising energy storage option. A renewable energy source-based hydrogen production and storage system is considered to be effective for energy management because renewable energy systems have intermittent characteristics [162].

Many substances found in nature contain hydrogen. Among them water, naturally found as brine (seawater), river water, rain, or well water, is the most abundant. Hydrogen can also be extracted from fossil hydrocarbons, biomass, hydrogen sulfide, and some other substances. When hydrogen is extracted from fossil hydrocarbon, all carbon dioxide must be processed (separated, sequestrated, etc.) such that no GHGs or other pollutants are emitted into the atmosphere and the hydrogen extraction process can then be called "green" [19].

Currently, 96 % of world hydrogen is produced using fossil fuels. Natural gas is the main raw material among them, and steam methane reforming (SMR) is the method most used. The main alternative methods for hydrogen production using various energy sources are shown in Fig. 7.9 [163].

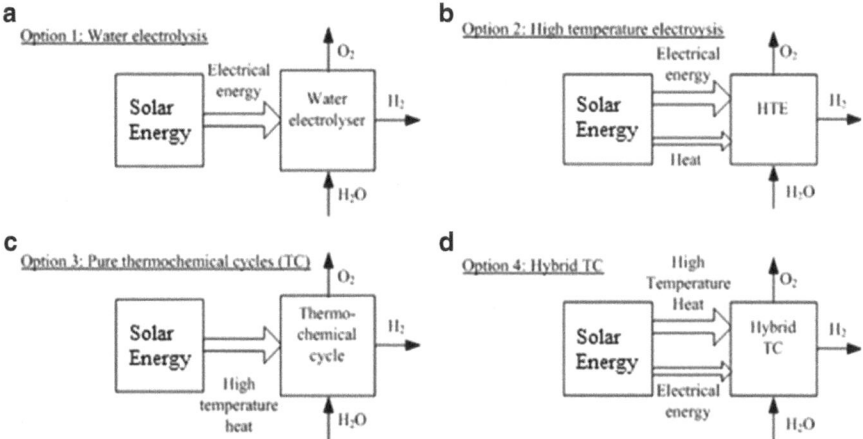

Fig. 7.10 Solar-based hydrogen production methods: (**a**) water electrolysis, (**b**) high-temperature electrolysis, (**c**) thermochemical cycles and (**d**) hybrid thermochemical cycles

The objective of this study is to investigate environmental impacts of solar-based hydrogen production via thermochemical water splitting using Mg-Cl and Cu-Cl cycles by performing life-cycle assessment (LCA). LCA is essentially a cradle-to-grave analysis to investigate environmental impacts of a system or process or product.

7.6.2 Solar-Assisted Hydrogen Production Methods Considered

Using solar energy as the primary energy source for hydrogen production is attractive because (i) the greenhouse gas emissions associated with solar energy production are more significantly reduced than with conventional fossil fuel combustion, and (ii) solar energy is adaptable to large-scale hydrogen production despite its intermittent characteristics. Significant progress in solar-based hydrogen production has been reported in recent years in the open literature. Solar-based hydrogen production methods, such as water electrolysis, high-temperature electrolysis, thermochemical water splitting, and hybrid thermochemical cycles for water splitting are promising (Fig. 7.10).

7.6.3 Life-Cycle Assessment and Process Analysis

Life-cycle assessment (LCA) is an analytical tool for evaluating the environmental impacts of the processes associated with goods and services and for identifying

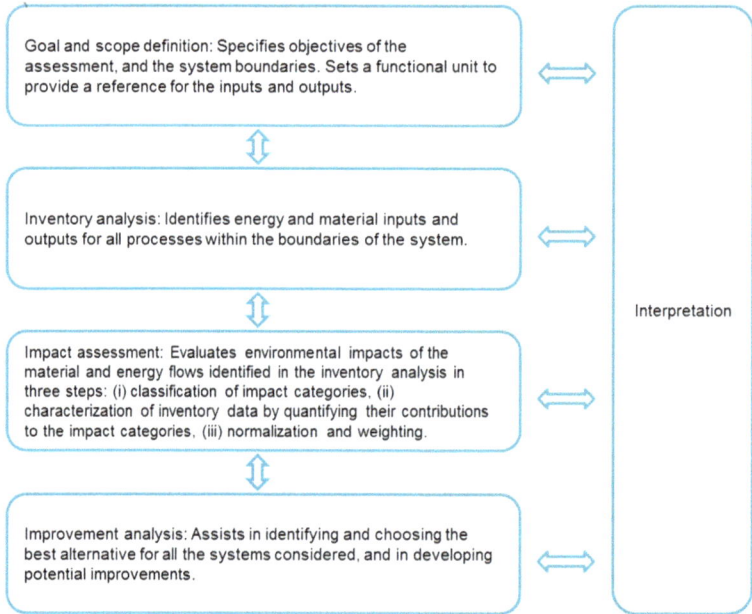

Fig. 7.11 The main steps of life-cycle assessment (LCA) and their descriptions

opportunities to reduce the impacts attributable to wastes and resource consumption. The tool considers all stages of a product or process, from production to disposal. LCA consists of four main phases (Fig. 7.11).

In this chapter, 1 kg hydrogen production is considered as the functional unit. Also, global warming potential (GWP) as the impact category is considered in the life-cycle impact assessment (LCIA) phase. GWP, which is defined as the impact of human emissions on the radiative forcing (i.e., thermal radiation absorption) of the atmosphere, and leads to climate change, which may affect ecosystem and human health. The earth's surface temperature is increased by emissions of greenhouse gases, which enhance radiative forcing (the "greenhouse effect"). GWP is measured in units of kg CO_2-eq.

7.6.4   Systems Description

The heliostat solar thermal tower and hydrogen production plant are two main subsystems of solar-based hydrogen production via thermochemical water splitting. A heliostat solar tower provides the energy requirement of the hydrogen production plant. Copper-chlorine (Cu-Cl) thermochemical water decomposition method is considered for the case study. Environmental impacts of 1 kg hydrogen produced using the solar-based Cu-Cl cycle are evaluated and compared with selected hydrogen production methods.

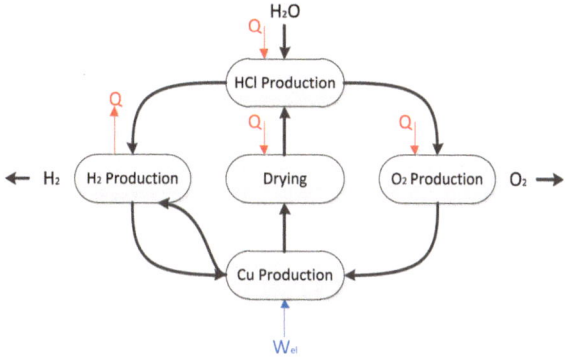

Fig. 7.12 Conceptual layout of the five-step Cu-Cl cycle

Cu-Cl thermochemical cycles involve a series of chemical reactions and are characterized by the number of major chemical steps they incorporate. Two-, three-, four-, and five-step Cu-Cl cycles for thermochemical water decomposition have been identified previously. The chemical reactions used in the Cu-Cl cycle, which utilizes a series of copper and chlorine compounds, form a closed loop, and all chemicals are recycled. These are participating chemicals that are relatively safe, inexpensive, and abundant. All chemical reactions in the Cu-Cl cycle are also proven in the laboratory with no significant side reactions [161]. Inputs to the Cu-Cl thermochemical water decomposition cycle are water, thermal energy, and electricity, with oxygen and hydrogen are the outputs.

This study considers three main types of Cu-Cl cycles for thermochemical water splitting (three-, four-, and five-step). Figure 7.12 shows a conceptual schematic of the five-step Cu-Cl cycle, which consists of the following main steps: (1) HCl (g) production, (2) oxygen production, (3) copper production, (4) drying, and (5) hydrogen production. The four-step cycle combines the third and fourth steps in the five-step cycle, whereas the three-step Cu-Cl cycle integrates the H_2 production step and the combined step in the four-step cycle to reduce the complexity and equipment requirements.

7.6.5 Analysis

The hydrogen production plant (construction and utilization) is modeled using GaBi 4 LCA software to investigate global warming potential. Inventory data on thermochemical hydrogen production are reported in the literature. Detailed information about the inventory data and the GaBi 4 model of the system and its analysis is presented elsewhere [164, 165]. The emission of greenhouse gases during the provision of services and the production of materials needed for the construction and operation of solar plant are also calculated. Greenhouse gas emissions associated with a solar tower are 32.7 (g CO_2-eq/kWh) [166].

The relationship between environmental impacts and economic aspects is also presented using the social cost of carbon concept. The social cost of carbon (SCC)

is the value of the climate change impacts from 1 tonne of carbon emitted today as CO_2, aggregated over time and discounted back to the present day. The average social cost of carbon is $160 per tonne of CO_2 emissions [159] (2007 U.S. dollars are used in this comparison).

In addition to an environmental impact assessment, an efficiency analysis is performed to determine the relationship between the LCA results and hydrogen plant efficiency. For the present analysis, hydrogen plant efficiencies are considered rather than the overall system efficiency, thereby excluding the energy source for hydrogen production. Hence, this efficiency is fixed as the electrolysis efficiency for hydrogen production via electrolysis using renewable energy (wind, solar, biomass). The energy efficiency is the ratio of the product output, hydrogen, which has a lower heating value (LHV) of 120 MJ/kg H_2, and the inputs (thermal energy and electrical energy). Thus, the efficiency of hydrogen plant can be calculated as follows:

$$\eta_{hydrogen\,plant} = \frac{LHV_{H_2}}{Q + W_{el}} \tag{7.1}$$

7.6.6 Results and Discussion

LCA results of the solar-based Cu-Cl cycles are presented and compared with selected hydrogen production methods: solar-based thermochemical Mg-Cl, solar-based thermochemical S-I cycle, solar PV-based electrolysis, nuclear-based ISPRA Mark 9 thermochemical cycle, nuclear-based high-temperature electrolysis, natural gas steam reforming, biomass-based electrolysis, and wind-based electrolysis [165, 167, 168]. Results presented here are for 1 kg hydrogen production.

Figure 7.13 compares hydrogen production via a solar-based thermochemical cycle with other methods in terms of (a) GWP and (b) SCC. Figure 7.13a shows that solar-based thermochemical cycles and wind-based electrolysis are the most environmentally benign hydrogen production methods. Natural gas steam reforming, on the other hand, has the highest environmental impact with respect to greenhouse gas emissions. GWP of the solar-based five-step Cu-Cl cycle is 1.1 kg CO_2-eq, whereas natural gas steam reforming has a GWP value of 11.89 kg CO_2-eq. Corresponding SCC values for the solar-assisted five-step Cu-Cl and natural gas steam reforming are, respectively, 18.85 and 190.21¢ per kg H_2 production.

Table 7.6 shows hydrogen plant efficiency values for selected methods. As stated in the analysis part, hydrogen plant efficiencies are considered rather than the overall system efficiency, thereby excluding the energy source for the hydrogen production. Hydrogen production using thermochemical water splitting via Cu-Cl cycles has energy efficiency values of approximately 50 %. The hydrogen production method with the highest efficiency values are natural gas steam reforming and electrolysis.

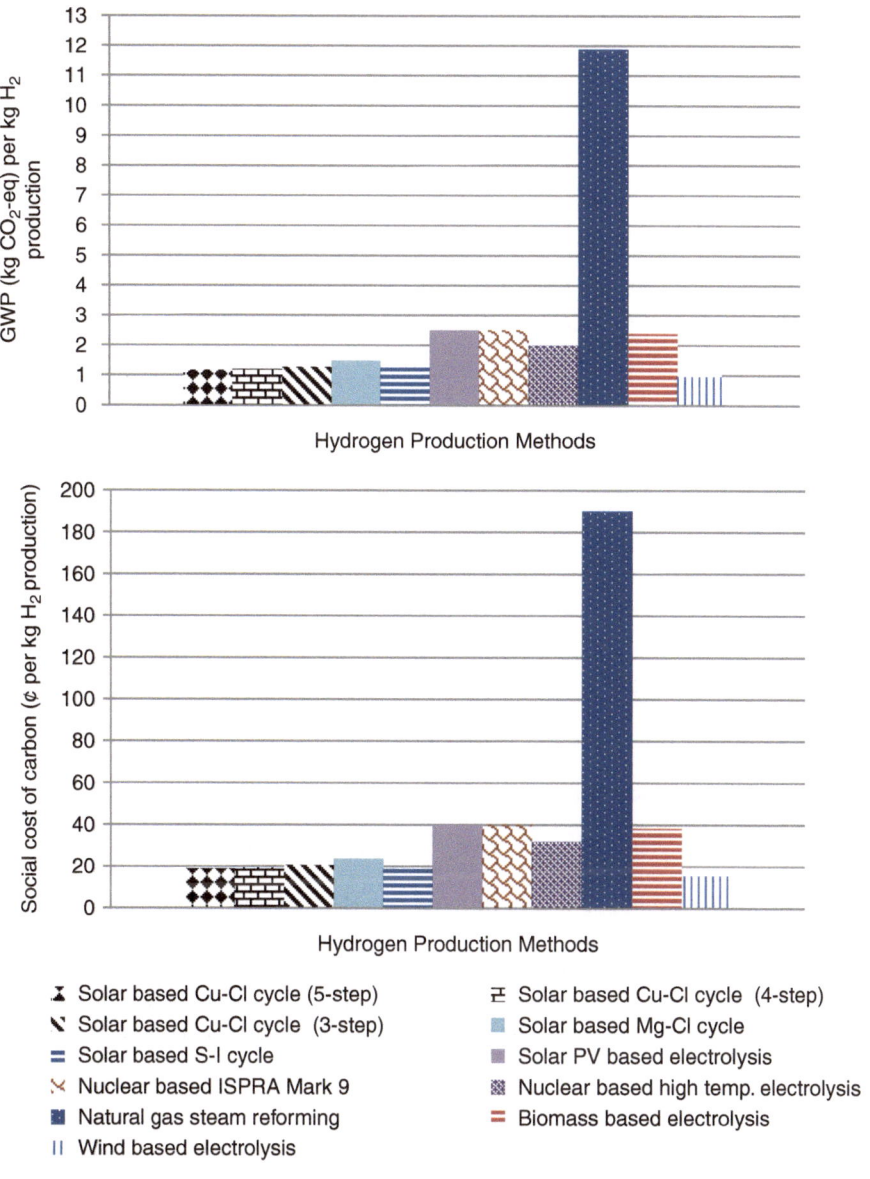

Fig. 7.13 Global warming potential (GWP) (*top*) and social cost of carbon (SCC) (*bottom*) for several hydrogen production methods (Modified from Ozbilen et al. [168])

Figure 7.14 presents the variation of hydrogen plant efficiency with energy requirements (heat and electrical work) of a hydrogen plant using thermochemical water splitting. The hydrogen plant efficiency is found using the efficiency equation. Figure 7.14 provides useful information regarding the hydrogen plant efficiency because the electrical work and heat requirements are dependent on the

Table 7.6 Energy efficiency values for several hydrogen production methods

Hydrogen production method	Efficiency (%)
Cu-Cl cycle (5-step)	48
Cu-Cl cycle (4-step)	53
Cu-Cl cycle (3-step)	49
S-I cycle	42
Mg-Cl cycle	63
ISPRA Mark 9	25
High-temperature electrolysis	51
Natural gas steam reforming	66
Electrolysis	70

Source: Ozbilen et al. [168]

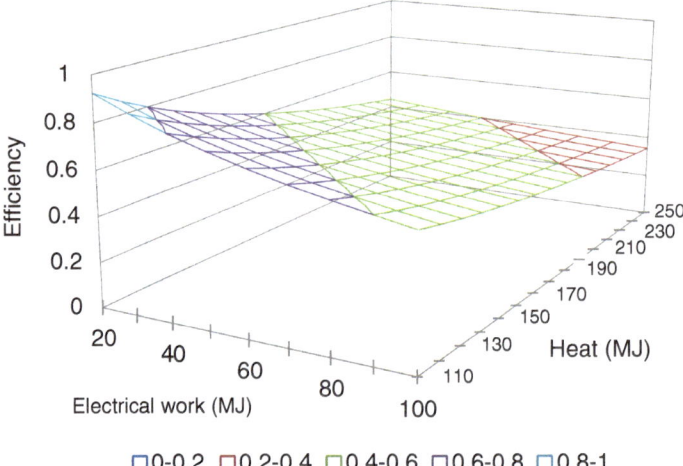

□0-0.2 □0.2-0.4 □0.4-0.6 □0.6-0.8 □0.8-1

Fig. 7.14 Variation of hydrogen plant efficiency with electrical work and heat (Modified from Ozbilen et al. [164])

number of steps of the thermochemical cycle. The energy efficiency approaches unity when both thermal and electrical energy requirements decrease.

7.6.7 Closing Remarks

Solar-based hydrogen production methods are reported, and environmental assessment is performed. The environmental impacts of various hydrogen production processes are evaluated and compared, considering several energy source using an LCA methodology. It is found that hydrogen produced by thermochemical water splitting cycles and wind-based electrolysis are more environmentally benign options compared to conventional natural gas steam reforming. Efficiency analysis shows that decreasing the energy requirements via thermal management increases hydrogen plant efficiency.

7.7 Case Study 6: Comparative Study of Integrated Solar Systems for Hydrogen Production

7.7.1 Introduction

Energy plays an important role in our daily life for almost all applications. At present, fossil fuels act as a base energy carrier and are treated as a damaging group for environment as they emit greenhouse gasses, such as CO_2, SO_2, and NOx.

Hydrogen is considered an environmentally benign energy carrier; however, it is not freely available in the atmosphere as compared to fossil fuels. Hydrogen can be produced using several techniques such as (a) steam methane reforming (SMR), (b) water electrolysis, (c) coal gasification, and (d) thermochemical cycles etc. Giaconia et al. [169] outlined that thermochemical water-splitting cycles (TWSCs) represent an appealing carbon-free option for hydrogen production powered by alternative (carbon-free) energy sources. Of many thermochemical cycles presented in the literature, the Cu-Cl cycle holds an edge in terms of temperature requirement and handling. The lower operating temperature requirements of Cu-Cl cycle (around 530 °C) have reduced material and maintenance costs as compared to other thermochemical cycles [170–173]. The major benefit of the Cu-Cl cycle is that it has the highest efficiency among all TWSCs; it can reach 55 %, as indicated by several researchers [e.g., 174, 175].

One of the very promising alternative and renewable sources of energy is solar energy. Solar energy converts solar flux reaching the Earth's surface into electricity or heat. The electricity generated by solar system is in direct current format, which can be used as it is or can be converted to alternative current based on end-user requirement. Major benefits of using a solar system include (a) environmentally benign operation, (b) no moving parts, (c) no wearing of parts if system is carefully protected from the environment, (d) energy output can vary from watts to megawatts based on the size of the system, (e) can be used to power phones or to power a community, and (f) module-by-module construction so that the size of the system can be altered based on the requirements. Integrated solar systems provide an attractive way of producing multiple outputs such as power, heat, hydrogen, and cooling in an environmentally benign manner. As it is expected that the future economy will be dominated by hydrogen fuel, many researchers have studied integrated solar systems for hydrogen production. Thomas and Nelson [176] stated that hydrogen fuel can be produced by using solar electric energy from photovoltaic (PV) modules for the electrolysis of water without emitting carbon dioxide or requiring fossil fuels. The results of analyses conducted by different researchers [e.g., 10, 177, 178] related to solar hydrogen production show that integrated solar production system is a very promising technology as it produces hydrogen in an environmental friendly and cost-effective manner. Also, studies have shown that using solar for hydrogen production enhances the efficiency of the overall system. In this case study, the aim is to perform energy and exergy analyses

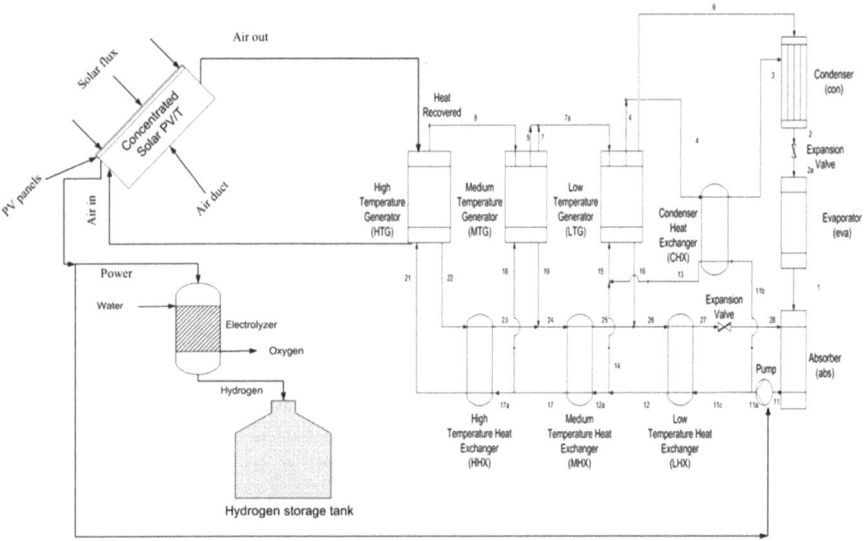

Fig. 7.15 Schematic of integrated solar system for hydrogen and cooling production

of integrated solar energy systems and to compare their performance based on energy and exergy efficiencies and energy required to produce a unit (kg/l) of hydrogen.

7.7.2   System Description

In this case study, we have studied two integrated systems, namely, (a) an integrated concentrated solar photovoltaic thermal (PV/T) absorption cooling system for the hydrogen and cooling production (Fig. 7.15) and (b) an integrated solar thermal, Cu-Cl cycle, and Kalina cycle for hydrogen production (Fig. 7.16).

In the first integrated system, concentrated solar PV/T is used to produce power and heat. Power produced is supplied to the electrolyzer and the pump in the cooling cycle. The electrolyzer is utilized to break the water molecule bond. As the water molecule breaks, it splits into hydrogen and oxygen. The hydrogen molecules are taken out of the electrolyzer and stored in a tank for later use as an energy source. The high-temperature air coming out of the solar PV/T is fed into the HTG of the absorption cooling system and is used as an energy source for the absorption cooling system. Detailed description of this system can be found in Ratlamwala et al. [178].

In the second integrated system, heat is produced using high-temperature solar thermal systems. The heat produced is supplied to the Cu-Cl cycle for hydrogen production (Fig. 7.17). A heat exchanger network for Cu-Cl cycle is also developed, which helps in better utilization of the heat within the cycle (Fig. 7.18).

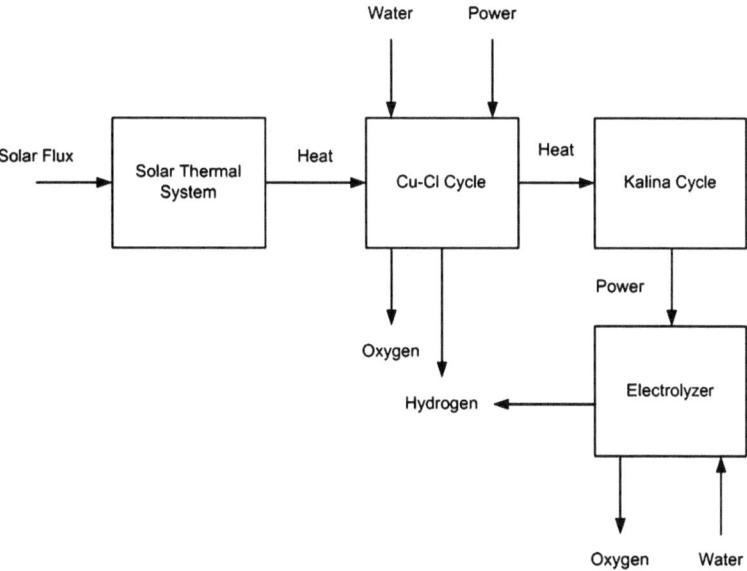

Fig. 7.16 Schematic of integrated solar thermal, Cu-Cl cycle, and Kalina cycle for hydrogen production

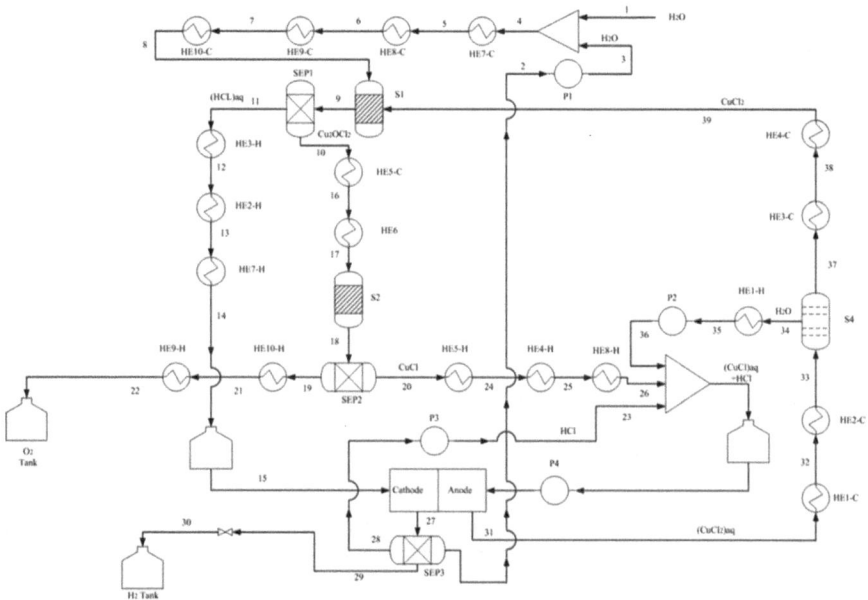

Fig. 7.17 Schematic diagram of Cu-Cl cycle

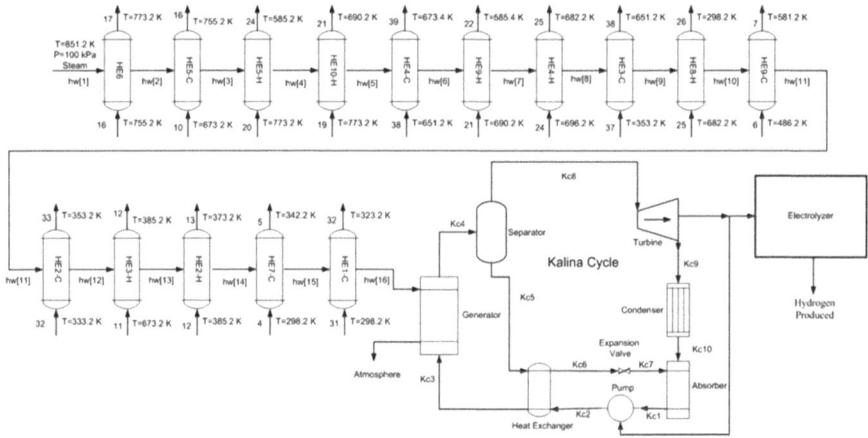

Fig. 7.18 Schematic diagram of heat exchanger network and Kalina cycle

The moderate temperature stream of water coming out of the heat exchanger network is then supplied to the Kalina cycle to produce power. The power produced by the Kalina cycle is later supplied to the electrolyzer to produce extra hydrogen. Detailed description of this system can be found elsewhere [179].

7.7.3   Energy and Exergy Analyses

The energy and exergy analyses of two systems studied in this case study are presented in this section.

For System 1:
The equation used to calculate power produced by the PV module is given as

$$\dot{W}_{solar} = \eta_c \times \dot{I} \times \beta_c \times \tau_g \times A \qquad (7.2)$$

The rate of heat transfer is given by

$$\dot{Q}_{solar} = \frac{\dot{m}_a \times cp_a}{U_L} \times \left(\left(h_{p2G} \times z \times \dot{I} \right) - U_L \times (T_{ai} - T_0) \right)$$
$$\times \left[1 - exp \left(\frac{-b \times U_L \times L}{\dot{m}_a \times cp_a} \right) \right] \qquad (7.3)$$

where

$$z = \alpha_b \times \tau_g^2 \times (1 - \beta_c) + h_{p1G} \times \tau_g \times \beta_c \times (\alpha_c - \eta_c)$$

The rate of exergy of solar energy is calculated by

$$\dot{E}_{x_{solar}} = \left[1 - \left(\frac{T_0 + 273.15}{T_{sun}}\right)\right] \times \dot{I} \times A \tag{7.4}$$

The electrical efficiency of the solar PV/T system is defined as

$$\eta_{el} = \eta_c \times (1 - 0.0045 \times (T_c - 25)) \tag{7.5}$$

where

$$T_c = \frac{\tau_g \times \beta_c \times \dot{I} \times (\alpha_c - \eta_c) + U_t \times T_0 \times h_t \times T_{bs}}{U_t + h_t}$$

$$T_{bs} = \frac{z \times \dot{I} + (U_t + U_{tb}) \times T_0 \times h_{ba} \times T_{air}}{U_b + h_{ba} + U_{tb}}$$

$$T_{air} = \left[T_0 + \frac{h_{pzG} \times z \times \dot{I}}{U_L}\right] \times \left[1 - \frac{1 - exp\left(\frac{-b \times U_L \times L}{\dot{m}_a \times cp_a}\right)}{b \times U_L \times L/\dot{m}_a \times cp_a}\right] + T_{ai}$$

$$\times \left[\frac{1 - exp\left(-b \times U_L \times L/\dot{m}_a \times cp_a\right)}{b \times U_L \times L/\dot{m}_a \times cp_a}\right]$$

The thermal efficiency of the solar PV/T system is defined as

$$\eta_{th} = \frac{\dot{Q}_{solar}}{\dot{I} \times b \times L} \tag{7.6}$$

The electrolyzer is used to split the water molecule into hydrogen and oxygen molecules, where the hydrogen molecule is stored in the tank for later usage. The amount of hydrogen produced depends on the efficiency of the electrolyzer, higher heating value of the hydrogen, and power input to the electrolyzer. The rate of hydrogen produced by electrolyzer can be defined as

$$\eta_{electrolyzer} = \frac{\dot{m}H_2 \times HHV}{\dot{W}_{solar} - \dot{W}_{pump}} \tag{7.7}$$

The amount of heat supplied to the HTG is given below

$$\dot{Q}_{HTG} = \dot{Q}_{solar} \tag{7.8}$$

The energy balance equation for High temperature heat exchanger (HHX) is given below

$$\dot{m}_{17a} h_{17a} + \dot{Q}_{HHX} = \dot{m}_{21} h_{21} \tag{7.9}$$

$$\dot{m}_{22} h_{22} + \dot{Q}_{HHX} = \dot{m}_{23} h_{23} \tag{7.10}$$

The mass and energy balance equations for the condenser are given below:

$$\dot{m}_2 = \dot{m}_6 + \dot{m}_3 \tag{7.11}$$

$$\dot{m}_2 h_2 + \dot{Q}_{con} = \dot{m}_6 h_6 + \dot{m}_3 h_3 \tag{7.12}$$

The following equation is for the energy balance of the evaporator:

$$\dot{m}_{2a} h_{2a} + \dot{Q}_{eva} = \dot{m}_1 h_1 \tag{7.13}$$

The following energy balance equation is used to calculate the heat rejected by the absorber.

$$\dot{m}_{11} h_{11} + \dot{Q}_{abs} = \dot{m}_1 h_1 + \dot{m}_{28} h_{28} \tag{7.14}$$

The thermal exergy of evaporator and HTG are defined as

$$\dot{E}x_{th, eva} = \left(1 - \frac{T_0}{T_{eva}}\right) \times \dot{Q}_{eva} \tag{7.15}$$

$$\dot{E}x_{th, HTG} = \left(1 - \frac{T_0}{T_{HTG}}\right) \times \dot{Q}_{HTG} \tag{7.16}$$

The energy and exergy COPs are calculated as

$$COP_{en} = \frac{\dot{Q}_{eva}}{\dot{Q}_{HTG} + \dot{W}P} \tag{7.17}$$

$$COP_{ex} = \frac{Ex_{th, eva}}{Ex_{th, HTG} + \dot{W}P} \tag{7.18}$$

The energy and exergy efficiencies are defined as

$$\eta_{en} = \left(\frac{\dot{m}H_2 \times HHV + \dot{Q}_{eva}}{I \times b \times L}\right) \times 100 \tag{7.19}$$

$$\eta_{ex} = \left(\frac{Ex_{H_2} \times Ex_{th, eva}}{Ex_{so}}\right) \times 100 \tag{7.20}$$

For System 2:
The specific enthalpy at any given state in the Cu-Cl cycle is calculated as

$$h_i = \sum_{m=1}^{k} mf_m h_m \tag{7.21}$$

where h_i represents specific enthalpy at any given state, and mf_m and h_m represent mass fraction and specific enthalpy of a compound, respectively.

The specific entropy at any given state in the Cu-Cl cycle is calculated as

$$S_i = \sum_{m=1}^{k} mf_m S_m \tag{7.22}$$

where S_i represents specific entropy at any given state, and mf_m and S_m represent mass fraction and specific entropy of a compound, respectively.

The mass fraction of a substance is found by

$$mf_m = y_m \frac{M_m}{M_i} \tag{7.23}$$

with

$$M_i = \sum_{m=1}^{k} y_m M_m$$

where y_m represents the molar fraction of a compound. M_m and M_i represent the molar weight of a compound and a solution, respectively.

The exergy rate at any given state in the Cu-Cl cycle is

$$\dot{E}x_i = \dot{m}_i((h_i - h_0) - T_0(S_i - S_0)) \tag{7.24}$$

where $\dot{E}x_i$, $\dot{m}_i$, h_0, T_0, and S_0 represent exergy rate at any given state, mass flow rate at any given state, specific enthalpy at ambient state, ambient temperature, and specific entropy at ambient state, respectively.

The thermal exergy rate in each heat exchanger is defined as

$$\dot{E}x_{th_i} = \left(1 - \frac{T_0}{T_i}\right)\dot{Q}_i \tag{7.25}$$

where $\dot{E}x_{th_i}$, T_i, and $\dot{Q}_i$ represent thermal exergy rate, temperature of a system, and heat rejected by the system, respectively.

The power required by any pump in the Cu-Cl cycle is written as

$$\dot{W}_{pi} = \dot{m}_i(h_i - h_{i-1}) \tag{7.26}$$

where $\dot{W}_{pi}$ represents power required by the pump. h_i and h_{i-1} denote specific enthalpy at the exit of the pump and specific enthalpy at the entrance of the pump, respectively.

The energy and exergy efficiencies of the integrated system are defined as

$$
\eta_{en} = \left(\frac{\dot{m}_{29}HHV_{H_2} + \dot{m}_{22}h_{22} + (\dot{m}H_2HHV_{H_2})_{elec}}{\dot{Q}HE_3 - C + \dot{Q}HE_{10} - C + \dot{Q}S_1 + \dot{Q}S_2 \atop + \dot{Q}S_4 + \dot{m}_1h_1 + \dot{m}_{nw}\left(h_{nw[1]} - h_{nw[0]}\right) + \frac{(W_{elec}+W_{P1}+W_{P2}+W_{P3}+W_{P4})}{0.4}} \right)
$$

(7.27)

and

$$
\eta_{ex} = \left(\frac{\dot{E}xH_2 + \dot{E}x_{32} + \dot{E}xH_{2,\,elec}}{\dot{E}xHE_3 - C + \dot{E}xHE_{10} - C + \dot{E}xS_1 + \dot{E}xS_2 \atop + \dot{E}xS_4 + \dot{E}x_1 + \dot{E}x_{nw[1]} + \frac{(W_{elec}+W_{P1}+W_{P2}+W_{P3}+W_{P4})}{0.4}} \right)
$$

(7.28)

7.7.4   Results and Discussion

The results of two systems studied, namely, (a) a solar PV/T integrated with triple effect absorption cooling system and electrolyzer (system 1) and (b) a solar thermal Cu-Cl integrated with Kalina cycle (system 2) are presented in this section. The performances of these two systems are compared on the basis of rate of hydrogen production, energy and exergy efficiencies, and energy required to produce hydrogen in units of l/s (liters per second). The effect of rise in ambient temperature on performances of these five systems is also presented.

The effect of rise in ambient temperature on the hydrogen production rate of two systems studied is presented in Fig. 7.19. It is noticed that the rate of hydrogen production of system 1 and system 2 increases from 2.73 l/s to 3.23 l/s and 6.38 l/s to 7.54 l/s, respectively, with rise in ambient temperature from 273 to 323 K. It is observed that system 2 produced a higher rate of hydrogen as compared to system 1. The rate of hydrogen production should not be the only criterion when selecting the hydrogen production system. As seen in Fig. 7.20, the energy efficiency of the systems studied shows a completely different picture as to which system is producing hydrogen in the most efficient manner. The analysis shows that the energy efficiency of system 1 and system 2 remains constant at approximately 14 % and 59 %, respectively, with rise in ambient temperature. The first thing noticed is that the energy efficiency of these systems does not vary with rise in ambient temperature, and this behavior is noticeable because energy efficiency does not take into account the losses that occur from the temperature difference between the system and its surroundings. The energy efficiency analysis show that between the two systems studied, system 2 performs best. This trend is observed because thermochemical hydrogen production systems are very efficient as compared to other

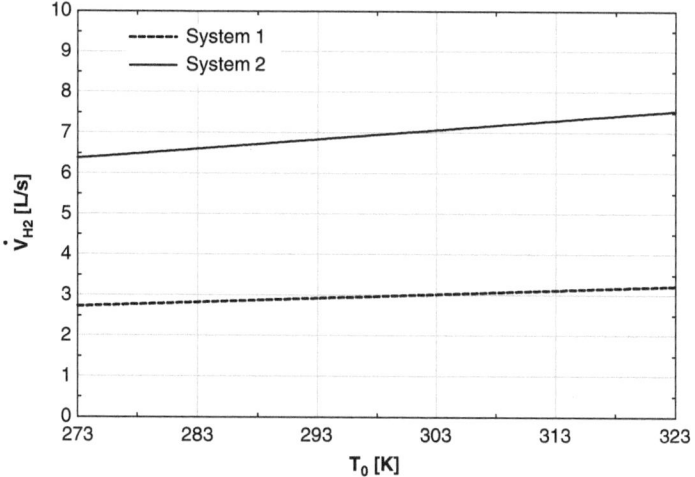

Fig. 7.19   Effect of rise in ambient temperature on rates of hydrogen production

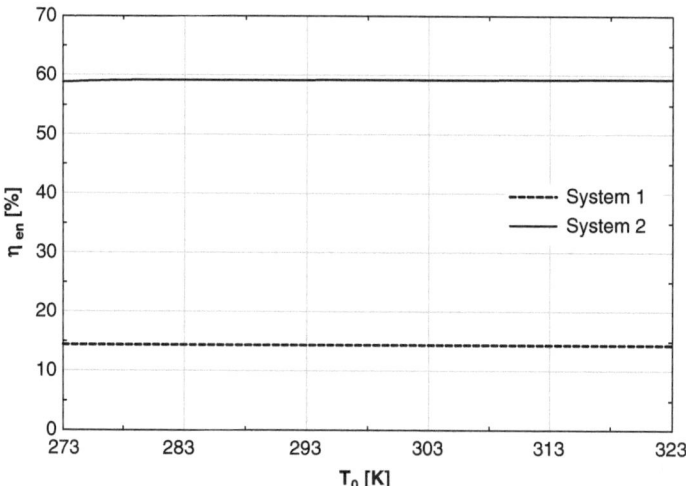

Fig. 7.20   Effect of rise in ambient temperature on energy efficiencies

systems as thermochemical systems mostly rely on heat to produce hydrogen as compared to power for other systems using a water electrolyzer. The power requirement of the electrolyzer of the thermochemical system is generally lower as compared to the water electrolyzer, because electrolysis of the thermochemical system is provided with chemical solutions (such as aqueous CuCl and HCL) in addition to water to produce hydrogen, which helps lower the voltage requirement of the electrolyzer. The effect of rise in ambient temperature on the exergy efficiencies of the systems studied is presented in Fig. 7.21. The exergy efficiency

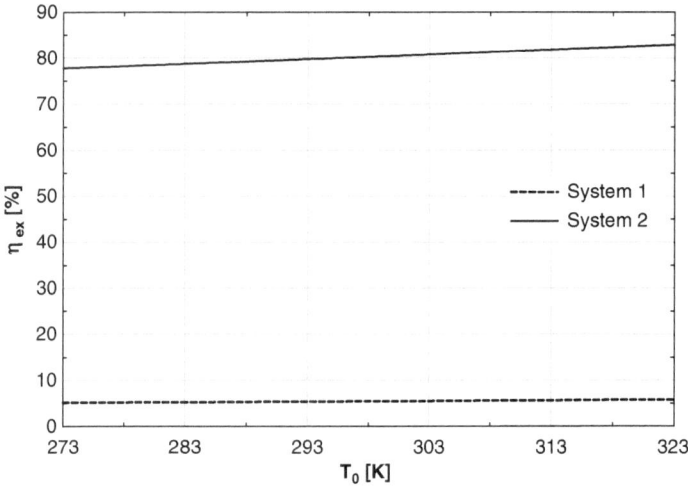

Fig. 7.21 Effect of rise in ambient temperature on exergy efficiencies

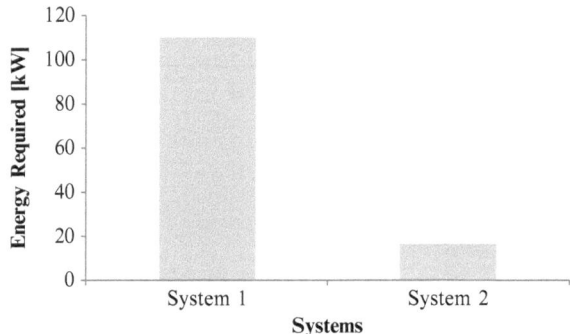

Fig. 7.22 Energy required by each system to produce unit l/s [liters (L) per second] of hydrogen

of system 1 and system 2 is found to be increasing from 5 % to 6 %, and from 78 % to 83 %, respectively, with rise in ambient temperature. The exergy efficiency is found to be increasing because rise in ambient temperature results in lower heat loss to the surroundings from the system as there is a lower temperature difference between the system and its surroundings. As losses occurring from heat transfer from the system to the surroundings decrease, the performance of the system increases. It is observed that from the exergy perspective system 2 performs better than system 1. Such a trend is expected because thermochemical cycles recover a vast amount of heat within the system whereas the source of solar systems is the sun, which contains an ample amount of energy that cannot be harnessed by the presently available solar systems as a result of material constraints. The amount of energy required to produce unit l/s of hydrogen by each system (Fig. 7.22) indicates that system 2 requires a lesser amount of energy (16.58 kW) to produce unit l/s of hydrogen as compared to system 1 (110.1 kW). This bar chart shows that the Cu-Cl cycle is the best system based on exergy parameters and energy required to produce unit l/s of hydrogen among the systems studied here.

7.7.5 Closing Remarks

This case study has presented a comparative study of the two solar-based hydrogen production systems to highlight which system performs best in terms of energy and exergy efficiencies and hydrogen production rates. The present systems are (a) solar PV/T integrated with triple effect absorption cooling system and electrolyzer (system 1) and (b) solar thermal Cu-Cl integrated with Kalina cycle (system 2). Although the energy efficiencies of both systems are not greatly affected by ambient temperature, the exergy efficiencies of these systems are found to increase with rise in ambient temperature. The study also shows that the energy required by system 1 to produce unit l/s of hydrogen is far greater than is required by system 2. This study concludes that system 2 performed better between the two systems studied.

Chapter 8
Concluding Remarks

The present book on solar hydrogen production covers almost all the available solar hydrogen production techniques in terms of basic understanding of their concepts, different systems and system components, various chemical reactions and methods, thermodynamic analysis based on energy and exergy efficiencies, sustainability, and environmental impact. A new concept based on solar hydrogen production via artificial photosynthesis is also discussed. Some case studies are also presented to implement the analysis part with some actual data. Some appropriate concluding remarks are as follows.

- Hydrogen energy is capable of replacing fossil fuels completely in the different energy-intensive sectors, such as transportation, commercial and industrial, domestic, and agricultural. Hydrogen can totally fulfill the energy needs of urban and rural areas and provide a better alternative for fuel in automobiles and other vehicles also.
- Hydrogen is naturally present on Earth in a combined state in both organic and inorganic compounds, but it is rarely present in the free and molecular state. Therefore, elemental hydrogen must be artificially produced, and thus its safe and environmentally benign production is most important.
- In many available hydrogen production technologies, solar hydrogen production technology is quite fascinating as the input energy source; that is, solar energy is freely and abundantly available, and its environmental impact is much less as solar is a sustainable source of energy.
- Solar thermal hydrogen production and photochemical and PV-based hydrogen production methods are well-established methods whereas intensive research is needed to improve the performance of biological hydrogen production and bio-photolysis methods.
- Solar thermal technology has the capability to produce hydrogen for different temperature ranges (200–2,000 °C) by using different chemical reactions. Solar thermolysis and high-temperature electrolysis are high temperature, whereas solar thermochemical cycles, solar gasification, reforming, and cracking are medium- and high-temperature technologies. Photochemical hydrogen

I. Dincer and A.S. Joshi, *Solar Based Hydrogen Production Systems*,
SpringerBriefs in Energy, DOI 10.1007/978-1-4614-7431-9_8, © The Author(s) 2013

conversion and low-temperature electrolysis come under the category of low-temperature technology.

- Most of the hydrogen production methods break water into hydrogen and oxygen; therefore, water is the main reactant in almost all hydrogen production reactions.
- Electrolysis of water is the most common production method, whereas steam methane reforming is a commercially accepted method to produce hydrogen.
- Similar to the side effects of every medical prescription, every technology has some drawbacks. For example, intermittency and variance in the intensity of solar radiation handicap its use, but by using a thermal storage system and concentrating collectors, those drawbacks can be overcome. Therefore, it can be said that more research is needed to overcome the drawbacks of different technologies.
- The exergy efficiency of the concentrating collector is an important parameter to evaluate the exergy efficiency of the solar thermal hydrogen production system as it fluctuates with the intensity of solar radiation. Similarly, the exergy efficiency of a photovoltaic (PV) panel also fluctuates with the intensity of solar radiation and ambient temperature, therefore making it a critical parameter for the exergy efficiency calculation of the PV hydrogen production system. Design modifications and proper implementation and use of respective technologies may enhance the exergy efficiency of the system.
- The higher the exergy efficiency of the system, the higher would be the sustainability index. Solar thermal hydrogen systems have a higher sustainability index because of their higher exergy efficiency as compared to the photovoltaic hydrogen production system.
- Renewable energy sources (such as solar energy, wind, and geothermal) are less polluting as compared to conventional energy sources (particularly fossil fuels). In other words, the environmental impact of renewable energy sources is much less as compared to conventional energy sources, and therefore the renewable sources should be a preferred option for input energy for hydrogen production.
- By using renewable energy sources, not only can the environmental impact be reduced by a considerable reduction in greenhouse gas (GHG) release, but also the remaining fossil fuels can be conserved for future use.
- Energy recovery is equally important as compared to energy conversion because it can also be reused to produce hydrogen.

Nomenclature

A	Area (m^2)
A_{Apr}	Area of aperture (m^2)
A_S	Surface area of insulator (m^2)
b	Breadth of PV module (m)
C	Concentration factor
COP	Co efficient of performance
D_P	Depletion factor
DNI	Direct normal irradiance (W/m^2)
E	Energy generated by the fuel (J)
$\dot{E}n$	Energy rate (W/m^2)
$\dot{E}x$	Exergy destruction rate, Exergy rate (W/m^2)
FF	Fill factor
F_R	Efficiency factor
F_{Rec}	Efficiency factor of receiver
H	Specific enthalpy, kJ/kg; Heat transfer coefficient, W/m^{2}K
HHV_{H_2}	Higher heating value of Hydrogen, 39.4 kWh/kg
HHV	Higher heating value
Hw	High temperature water
I	Solar flux (W/m^2), Current (A)
I_{HB}	Terrestrial beam radiation (W/m^2)
I_{HD}	Terrestrial diffuse radiation (W/m^2)
I_N	Normal terrestrial solar radiation (W/m^2)
I_{ON}	Normal extraterrestrial solar radiation (W/m^2)
I_{SC}	Solar constant (1367 W/m^2)
I_t	Total solar radiation (W/m^2)

I. Dincer and A.S. Joshi, *Solar Based Hydrogen Production Systems*,
SpringerBriefs in Energy, DOI 10.1007/978-1-4614-7431-9, © The Author(s) 2013

k_1	Thermal conductivity of rector wall (W/m^{2}K)
k_2	Thermal conductivity of insulator (W/m^{2}K)
L	Length of the PV module (m)
m	Air mass (dimensionless)
M	Mass of the fuel (kg)
$\dot{m}$	Mass flow rate (kg/s)
$\mathbf{MW}$	Molecular weight, kg/mol
mf	Mass fraction
n	Day of the year (example: $n = 15$ for 15th January)
P	Power (W), Pressure, kPa
$\dot{Q}$	Thermal energy (W)
$\dot{Q}_{loss}1$	Conductive and convective heat loss through cylindrical surface (W)
$\dot{Q}_{loss}2$	Radiative heat loss through aperture (W)
$\dot{Q}_{loss}3$	Conductive and convective heat loss through back surface at closed end (W)
$\dot{Q}_{Solar}$	Solar energy (J)
$\dot{Q}_{Heliostats}$	Optical loss of the heliostat field (W)
$\dot{Q}_{Concentration}$	Optical loss at the receiver aperture (W)
$\dot{Q}_{Window}$	Optical loss the glass window (W)
$\dot{Q}_{Reradiation}$	Radiation emitted by the cavity
$\dot{Q}_{Wall}$	Conduction through the reactor walls (W)
$\dot{Q}_{Chemical}$	Energy transferred to the reactants and available for the reaction
r_1	Inner radius of cavity reactor (m)
r_2	Outer radius of cavity reactor (m)
r_3	Outer radius of insulator (m)
S	Specific entropy (kJ/kg.K)
S	Separator
SI	Sustainability index
$T_{Rec}T$	Temperature (°C)
T_a	Ambient temperature (K)
T_R	Cloudiness/haziness factor
T_{Rec}	Absorber/receiver surface temperature (K)
$\mathbf{TWSCs}$	Thermochemical water-splitting cycles
U_L	Heat transfer coefficient from absorber surface to ambient(W/m^{2}K)
U_C	Overall heat loss coefficient for cylindrical surface (W/m^{2}K)
U_B	Overall heat loss coefficient for back surface at closed end (W/m^{2}K)
V	Voltage (V); Volume of the fuel (Liter)
$\dot{W}$	Shaft work, Work rate (W)
Y	Mole fraction

Greek Letters

α	Lumped atmospheric parameter for beam radiation, Absorptivity
β	Surface tilt angle from horizontal (degree)
γ	Solar azimuth angle (degree)
δ	Angle of declination (degree)
ε	Optical thickness of the atmosphere, Emissivity
η	Efficiency, Energy efficiency, %
ρ	Reflectance, Reflectivity
τ	Torque, Nm, transmissivity of glass cover
ψ	Exergy efficiency, %
ω	Hour angle (degree), Rotations per minute (rpm)
σ	Stefan Boltzmann constant ($5.67 \times 10{-8}$ W/m^{2}K^4)
φ	Latitude of the place (degree)
θ_i	Angle of incidence of solar radiation (degree)
θ_Z	Zenith angle (degree)

Subscripts

a	Ambient
abs	Absorbed
A	Absorber
C	Collector
ch	Chemical
con	Condenser
CR	Charge regulator
D	Destruction
dc	Decomposer
EL, **elec**	Electrolyzer
E	Electrical
en	Energy
ex	Exergy
ext	Extra
GEN	Generator
g	*glass*
HE	Heat engine
H_2	Hydrogen
i	ith state
IN	Inverter
in	Input
k	kth state
M	Mechanical

m	mth state
mirror	Mirror
O_2	Oxygen
oc	Open circuit
out	Output
PV	Photovoltaic
p	Pump
ph	Physical
R	Reflector
s	Sun, Solar
sc	Short circuit
th	Thermal
turb	Turbine
u	Useful
1...25	State numbers
0	Ambient or reference condition

Appendix A

I. Dincer and A.S. Joshi, *Solar Based Hydrogen Production Systems*,
SpringerBriefs in Energy, DOI 10.1007/978-1-4614-7431-9, © The Author(s) 2013

Table A.1 Parameters on horizontal surface for sunshine hours $= 10$ for all four weather type of days for different Indian climates

Type of day	Parameters	January	February	March	April	May	June	July	August	September	October	November	December
1. Parameters on horizontal surface for sunshine hours $= 10$ for all four type of days for New Delhi
a	T_R	2.25	2.79	2.85	2.72	3.54	2.47	2.73	2.58	2.53	1.38	0.62	0.72
	α	0.07	0.10	0.17	0.23	0.16	0.28	0.37	0.41	0.29	0.47	0.59	0.54
	K_1	0.47	0.39	0.33	0.28	0.20	0.27	0.41	0.40	0.23	0.21	0.21	0.28
	K_2	−13.17	−6.25	5.61	38.32	65.04	31.86	−40.57	−55.08	39.92	32.77	30.62	9.73
b	T_R	2.28	2.78	2.89	3.15	5.44	4.72	5.58	5.43	3.23	4.56	0.19	1.83
	α	0.15	0.13	0.14	0.17	0.16	0.20	0.24	0.18	0.31	0.22	1.14	0.42
	K_1	0.51	0.54	0.49	0.46	0.45	0.45	0.53	0.39	0.37	0.42	0.35	0.40
	K_2	−21.77	−28.26	−9.22	−11.55	1.54	23.99	−51.61	9.46	14.07	−9.50	17.47	−0.07
c	T_R	5.88	6.36	6.11	7.77	9.20	10.54	7.13	7.97	5.51	5.01	4.93	3.23
	α	0.27	0.37	0.37	0.31	0.07	0.06	0.41	0.51	0.49	1.26	1.06	0.64
	K_1	0.39	0.36	0.33	0.35	0.56	0.48	0.47	0.35	0.39	0.36	0.31	0.43
	K_2	−14.73	−7.97	10.87	20.45	−56.00	−0.37	−52.27	47.70	35.64	−0.68	13.06	−7.04
d	T_R	7.47	8.97	10.77	11.18	13.69	12.47	8.21	8.58	9.40	7.24	4.30	4.02
	α	0.96	1.04	0.24	0.07	0.07	0.61	1.26	1.10	0.84	1.29	1.43	1.70
	K_1	0.35	0.30	0.43	0.49	0.48	0.46	0.43	0.43	0.41	0.36	0.31	0.38
	K_2	−25.89	−6.48	−36.46	−44.07	−42.58	−62.66	−56.75	−61.08	−27.09	3.90	20.10	−11.78
2. Parameters on horizontal surface for sunshine hours $= 10$ for all four type of days for Bangalore
a	T_R	3.36	3.27	3.63	5.05	4.24	4.32	5.18	4.75	4.10	2.28	1.66	1.65
	α	0.07	0.13	0.06	−0.06	0.10	0.19	0.10	0.18	0.13	0.33	0.35	0.36
	K_1	0.33	0.35	0.33	0.29	0.21	0.25	0.32	0.23	0.20	0.05	0.03	0.12
	K_2	−18.05	−22.11	−5.44	14.54	47.81	22.40	−26.04	10.14	38.54	107.04	103.64	47.70
b	T_R	3.24	5.25	6.21	5.72	5.90	7.35	4.12	5.27	4.83	2.43	1.89	3.68
	α	0.31	0.24	0.21	0.19	0.25	0.17	0.51	0.44	0.62	0.56	0.78	0.39
	K_1	0.50	0.45	0.48	0.50	0.41	0.50	0.46	0.50	0.33	0.26	0.37	0.41
	K_2	−60.12	−60.50	−80.04	−75.59	−28.55	−103.35	−90.54	−115.27	13.80	69.14	9.08	−33.76
c	T_R	3.70	4.51	7.74	5.83	4.95	4.39	5.68	2.67	6.64	4.71	5.68	2.02
	α	0.96	0.94	0.63	0.98	0.96	1.12	1.07	1.35	0.78	1.03	0.93	1.44

		1	2	3	4	5	6	7	8	9	10	11	12
d	K_1	0.43	0.36	0.43	0.48	0.55	0.50	0.58	0.53	0.50	0.36	0.57	0.46
	K_2	−47.21	−15.95	−26.53	−52.93	−161.61	−108.34	−156.14	−103.14	−61.13	−20.76	−129.68	−63.02
	T_R	2.80	3.84	3.91	3.94	6.68	4.45	4.84	6.33	6.86	7.35	7.49	6.13
	α	2.58	2.04	2.00	2.16	1.69	2.32	2.00	1.59	1.48	1.41	1.31	1.61
	K_1	0.27	0.55	0.42	0.38	0.50	0.41	0.61	0.53	0.45	0.40	0.30	0.29
	K_2	−12.29	−177.28	−125.35	−88.62	−146.94	−79.79	−213.29	−99.52	−72.22	−39.85	83.73	36.80

3. Parameters on horizontal surface for sunshine hours = 10 for all four type of days for Jodhpur

		1	2	3	4	5	6	7	8	9	10	11	12
a	T_R	1.54	1.56	2.26	3.20	3.39	3.25	3.87	3.72	2.82	1.59	1.33	1.26
	α	0.31	0.39	0.33	0.27	0.28	0.27	0.21	0.21	0.27	0.37	0.38	0.37
	K_1	0.26	0.23	0.24	0.26	0.17	0.10	0.13	0.20	0.21	0.18	0.14	0.22
	K_2	9.48	22.71	27.40	14.42	59.41	105.23	87.88	50.84	47.66	56.40	63.90	30.67
b	T_R	3.43	2.28	2.90	3.81	4.73	5.07	5.50	5.21	4.07	3.00	2.03	2.34
	α	0.24	0.46	0.38	0.35	0.40	0.37	0.28	0.23	0.31	0.42	0.55	0.46
	K_1	0.40	0.33	0.30	0.34	0.33	0.34	0.33	0.33	0.34	0.31	0.29	0.33
	K_2	−11.64	12.35	24.12	8.71	29.57	35.81	33.40	31.22	23.50	42.22	43.13	12.89
c	T_R	4.23	3.28	3.71	3.40	5.10	4.90	5.58	6.87	4.97	4.04	4.78	3.81
	α	1.06	1.31	2.05	0.97	0.88	1.02	0.67	0.61	0.64	0.98	1.32	0.93
	K_1	0.44	0.44	0.53	0.48	0.50	0.41	0.46	0.47	0.47	0.42	0.40	0.43
	K_2	−32.84	−35.85	−62.06	−26.96	−60.42	2.06	−35.15	−44.76	−26.93	−19.11	12.44	−33.72
d	T_R	1.94	1.71	7.67	1.63	3.17	9.62	3.52	8.01	9.33	7.09	5.20	2.25
	α	2.03	2.89	0.86	3.24	2.77	2.37	2.37	1.66	1.59	2.03	1.64	1.89
	K_1	0.39	0.36	0.52	0.44	0.44	0.52	0.28	0.43	0.44	0.42	0.46	0.44
	K_2	−14.88	−15.46	−26.47	−77.55	−87.34	−221.29	60.69	−117.01	−149.27	−89.92	−45.44	−19.31

4. Parameters on horizontal surface for sunshine hours = 10 for all four type of days for Mumbai

		1	2	3	4	5	6	7	8	9	10	11	12
a	T_R	3.27	2.97	3.16	4.22	4.25	3.31	3.20	5.40	3.95	2.88	1.80	1.95
	α	0.18	0.23	0.30	0.15	0.33	0.61	0.16	−0.02	0.14	0.23	0.37	0.34
	K_1	0.30	0.26	0.24	0.24	0.12	0.09	0.25	0.28	0.34	0.28	0.19	0.26
	K_2	−4.81	9.11	15.87	30.02	47.27	27.28	4.55	30.06	−0.75	27.13	53.96	19.77
b	T_R	4.21	3.40	3.93	4.78	6.70	7.74	6.08	6.25	4.98	3.57	2.68	2.96
	α	0.24	0.45	0.47	0.47	0.37	0.20	0.19	0.15	0.25	0.37	0.49	0.43

(continued)

Table A.1 (continued)

Type of day	Parameters ▼	January	February	March	April	May	June	July	August	September	October	November	December
	K_1	0.35	0.31	0.35	0.40	0.42	0.44	0.31	0.39	0.41	0.36	0.34	0.37
	K_2	−0.14	24.17	11.73	−13.57	−13.69	−19.52	61.35	22.16	−14.71	5.99	0.60	−14.17
c	T_R	3.06	2.26	3.24	4.39	5.91	5.97	8.17	4.24	5.36	3.16	2.97	3.75
	α	1.14	1.18	1.10	1.00	0.79	0.86	0.62	1.26	0.98	1.13	1.10	0.91
	K_1	0.59	0.58	0.52	0.54	0.60	0.52	0.54	0.43	0.44	0.47	0.57	0.54
	K_2	−59.86	−47.12	−58.09	−78.37	−111.97	−81.79	−95.21	−34.40	−39.31	−28.02	−48.41	−52.45
d	T_R	3.38	7.42	4.45	2.30	4.71	4.71	6.41	7.40	7.46	3.22	5.13	3.05
	α	1.71	1.73	2.29	2.08	2.95	2.66	2.68	1.81	2.14	2.15	1.53	1.51
	K_1	0.52	0.56	0.50	0.35	0.41	0.38	0.32	0.47	0.34	0.42	0.57	0.53
	K_2	−59.78	−26.16	−82.34	63.52	−101.81	−87.19	−61.50	−108.37	−38.68	−25.89	−78.03	−40.51

5. Parameters of on horizontal surface for sunshine hours = 10 for all four type of days for Srinagar

Type of day	Parameters ▼	January	February	March	April	May	June	July	August	September	October	November	December
a	T_R	1.45	5.37	3.31	4.25	5.41	3.63	5.77	6.45	4.06	2.61	4.03	0.72
	α	0.33	−0.36	−0.03	−0.03	−0.12	0.08	−0.09	−0.23	0.03	0.20	−0.37	0.53
	K_1	0.37	0.63	0.69	0.37	0.51	0.33	0.17	0.37	0.46	0.43	0.66	0.33
	K_2	−6.14	−82.86	−94.01	−10.95	−79.57	−13.73	68.06	−42.79	−60.27	−47.83	−37.00	−6.60
b	T_R	3.09	6.98	4.65	6.92	5.86	6.82	7.40	7.58	6.41	4.04	0.04	0.35
	α	0.38	−0.48	0.23	0.06	0.29	0.11	0.00	−0.13	−0.04	0.19	1.16	1.00
	K_1	0.39	0.83	0.59	0.42	0.32	0.63	0.48	0.38	0.48	0.52	0.37	0.41
	K_2	−23.08	−110.23	−107.74	−49.61	0.26	−167.86	−80.06	−13.91	−66.64	−62.52	−14.63	−12.20
c	T_R	2.35	6.59	6.31	7.57	8.69	8.00	9.72	8.23	7.36	5.02	1.86	0.76
	α	1.64	0.86	1.35	0.57	0.61	0.81	0.69	0.90	0.99	1.49	1.47	1.98
	K_1	0.41	0.42	0.48	0.54	0.50	0.39	0.56	0.49	0.44	0.52	0.41	0.31
	K_2	−37.87	−85.68	−180.45	−120.38	−146.97	−87.44	−228.91	−147.96	−62.10	−93.64	−40.07	−12.15
d	T_R	1.69	1.36	7.52	9.09	9.48	10.79	10.93	8.54	8.16	7.75	3.78	2.44
	α	2.63	2.97	1.87	1.35	1.13	1.56	3.08	1.71	3.15	1.70	1.74	2.04
	K_1	0.43	0.36	0.35	0.62	0.92	0.80	0.45	0.75	0.67	0.55	0.48	0.63
	K_2	−41.27	−44.68	−65.17	−254.24	−467.30	−421.63	−129.49	−356.92	−261.85	−119.53	−49.16	−64.02

Appendix B

I. Dincer and A.S. Joshi, *Solar Based Hydrogen Production Systems*,
SpringerBriefs in Energy, DOI 10.1007/978-1-4614-7431-9, © The Author(s) 2013

Table B.1 Number of days fall in different weather condition

Type of days	January	February	March	April	May	June	July	August	September	October	November	December
i. Bangalore												
a	7	8	9	9	7	8	7	6	8	7	6	7
b	16	11	11	10	13	11	13	11	10	10	10	10
c	6	7	8	9	8	8	7	10	9	11	12	12
d	2	2	3	2	3	3	4	4	3	3	2	2
ii. Jodhpur												
a	7	9	11	11	9	7	6	5	5	6	5	6
b	14	13	13	15	17	18	16	18	17	13	11	14
c	8	5	6	3	4	4	6	5	6	10	11	9
d	2	2	1	1	1	1	3	3	2	2	3	2
iii. Mumbai												
a	5	6	9	9	7	6	5	5	6	6	6	5
b	15	11	11	12	15	16	12	10	9	12	10	13
c	10	10	9	8	8	3	6	10	10	10	11	11
d	1	1	2	1	1	5	8	6	5	3	3	2
iv. New Delhi												
a	3	3	5	4	4	3	2	2	7	5	6	3
b	8	4	6	7	9	4	3	3	3	10	10	7
c	11	12	12	14	12	14	10	7	10	13	12	13
d	9	9	8	5	6	9	17	19	10	3	2	8
v. Srinagar												
a	5	7	8	10	13	11	7	6	14	12	5	3
b	17	14	17	17	15	12	17	18	12	12	14	8
c	7	4	3	2	2	4	4	3	3	5	8	19
d	2	3	3	1	1	3	3	4	2	2	3	1

References

1. Markandya A, Wilkinson P (2007) Electricity generation and health. Lancet 370 (9591):979–990
2. Veziroglu TN, Barbir F (1992) Hydrogen: the wonder fuel. Int J Hydrog Energy 17 (6):391–404
3. Veziroglu TN, Barbir F (1995) Transportation fuel-hydrogen. Energy technology and the environment, 4th edn. Wiley Interscience, NewYork, pp 2712–2730
4. Veziroglu TN (1987) Hydrogen technology for energy needs of human settlements. Int J Hydrog Energy 12(2)
5. Veziroglu TN, Sahin S (2008) 21st century's energy: hydrogen energy system. Energy Convers Manag 49:1820–1831
6. Barbir F, Veziroglu TN, Plass HJ Jr (1990) Environmental damage due to fossil fuels use. Int J Hydrog Energy 15(10):739–749
7. Internet source: hydrogen fact sheet. The basics of hydrogen. url: http://www. getenergysmart.org/Files/HydrogenEducation/4TheBasicsofHydrogen.pdf. Accessed on 15 June 2011 at 11:55 am
8. Muradov NZ, Veziroglu TN (2008) Review: "Green" path from fossil-based to hydrogen economy: an overview of carbon-neutral technologies. Int J Hydrog Energy 33:6804–6839
9. Zamfirescu C, Dincer I (2008) Environmentally-benign hydrogen production from ammonia for vehicles. In proceedings of: GCGW 2008, Istanbul, Turkey
10. Yilanci A, Dincer I, Ozturk HK (2009) A review on solar-hydrogen/fuel cell hybrid energy systems for stationary applications. Prog Energy Combust Sci 35(3):231–244
11. Balta MT, Dincer I, Hepbasli A (2009) Thermodynamic assessment of geothermal energy use in hydrogen production. Int J Hydrog Energy 34(7):2925–2939
12. Abanades S, Flamant G (2006) Solar hydrogen production from the thermal splitting of methane in a high temperature solar chemical reactor. Sol Energy 80:1321–1332
13. Liu Q, Hong H, Yuan J, Jin H, Cai R (2009) Experimental investigation of hydrogen production integrated methanol steam reforming with middle-temperature solar thermal energy. Appl Energy 86:155–162
14. Z'Graggen A, Haueter P, Maag G, Vidal A, Romero M, Steinfeld A (2007) Hydrogen production by steam-gasification of petroleum coke using concentrated solar power—III. Reactor experimentation with slurry feeding. Int J Hydrog Energy 32:992–996
15. Charvin P, Abanades S, Lemort F, Flamant G (2008) Analysis of solar chemical processes for hydrogen production from water splitting thermochemical cycles. Energy Convers Manag 49:1547–1556
16. de Falco M, Barba D, Cosenza S, Iaquaniello G, Marrelli L (2008) Reformer and membrane modules plant powered by a nuclear reactor or by a solar heated molten salts assessment of the design variables and production cost evaluation. Int J Hydrog Energy 33:5326–5334

17. Ni M, Leung MKH, Leung DYC (2008) Energy and exergy analysis of hydrogen production by a proton exchange membrane (PEM) electrolyzer plant. Energy Convers Manag 49:2748–2756

18. Zedtwitz PV, Petrasch J, Trommer D, Steinfeld A (2006) Hydrogen production via the solar thermal decarbonization of fossil fuels. Sol Energy 80:1333–1337

19. Dincer I (2012) Green methods for hydrogen production. Int J Hydrog Energy 37:1954–1971

20. Brown J, Kjellstrom B, Zinko H (1989) Import of solar energy-import of solar energy produced hydrogen from the equatorial belt to Sweden, JSB Ingenjorsbyra AB Trosa, Sweden (in Swedish)

21. Friberg R (1993) A photovoltaic solar-hydrogen power plant for rural electrification in India. Part 1: a general survey of technologies applicable within the solar-hydrogen concept. Int J Hydrog Energy 18(10):853–882

22. Bertani R (2005) World geothermal power generation in the period 2001–2005. Geothermics 34:651–690

23. Hammons TJ (2004) Geothermal power generation worldwide: global perspective, technology, field experience, and research and development. Electr Power Compon Syst 32:529–553

24. Sigurvinsson J, Mansilla C, Arnason B, Bontemps A, Marechal A, Sigfusson TI (2006) Heat transfer problems for the production of hydrogen from geothermal energy. Energy Convers Manag 47:3543–3551

25. Sigurvinsson J, Mansilla C, Lovera P, Werkoff F (2007) Can high temperature steam electrolysis function with geothermal heat? Int J Hydrog Energy 32:1174–1182

26. Mansilla C, Sigurvinsson J, Bontemps A, Marechal A, Werkoff F (2007) Heat management for hydrogen production by high temperature steam electrolysis. Energy 32:423–430

27. Sorensen B (2004) Renewable energy, 3rd edn. Elsevier Academic Press, Burlington

28. Miyamoto K (ed) (1997) Renewable biological systems for alternative sustainable energy production (FAO Agricultural Services Bulletin—128). Food and Agriculture Organization of the United Nations, Rome

29. Abudala A, Dincer I, Naterer G (2009) Exergy analysis of hydrogen production from biomass gasification. In international conference on hydrogen production, Osahawa, Ontario, Canada

30. de Sacramento EM, Sales AD, de Lima LC, Veziroglu TN (2008) A solar–wind hydrogen energy system for the Ceara´ state – Brazil. Int J Hydrog Energy 33:5304–5311

31. Sopian K, Ibrahim MZ, Daud WRW, Othman MY, Yatim B, Amin N (2009) Performance of a PV–wind hybrid system for hydrogen production. Renew Energy 34:1973–1978

32. Singh HN, Tiwari GN (2005) Evaluation of cloudiness/haziness factor for composite climate. Energy 20:1589

33. Kasten F (1965) A new table and approximate formula for relative optical air mass. J Arch Meteorol Geophy Bioklimatel Ser B 14:206–223

34. Kasten F, Young AT (1989) Revised optical air mass tables and approximation formula. J Appl Opt 28:4735–4738

35. Duffie JA, Beckman WA (1991) Solar engineering of thermal processes (2nd edn). Wiley Interscience, New York, pp 1–14

36. Liu BYH, Jordan RC (1960) The interrelationship and characteristic distribution of direct, diffuse and total solar radiation. J Sol Energy 4(3):1–19

37. Perez R (1992) Dynamic global-to-direct irradiance conversion models. J ASHRAE Trans 98 (1):354–369

38. Joshi AS, Tiwari GN (2005) Evaluation of cloudiness/haziness factor for composite climate of New Delhi. Solaris 2005, Athens, pp 116–120

39. Joshi AS (2006) Evaluation of cloudiness/haziness factor and its application for photovoltaic thermal (PV/T) system for Indian climatic conditions. Ph.D. thesis, IIT Delhi, New Delhi, India

40. Jing DW, Zhang YJ, Guo LJ (2005) Study on the synthesis of Ni doped mesoporous TiO_2 and its photocatalystic activity for hydrogen evolution in aqueous methanol solution. Chem Phys Lett 415:74–78

41. Liu MY, You WS, Lei ZB, Zhou GH, Yang JJ, Wu GP (2004) Water reduction and oxidation on Pt–Ru/Y2Ta2O5N2 catalyst under visible light irradiation. Chem Commun 19:2192–2193
42. Kato H, Asakura K, Kudo A (2003) Highly efficient water splitting into H_2 and O_2 over Lanthanum-doped NaTaO3 photocatalysts with high crystallinity and surface nanostructure. J Am Chem Soc 125:3082–3089
43. Yoon TP, Ischay MA, Juana D (2010) Visible light photocatalysis as a greener approach to photochemical synthesis. Nat Chem 2(7):527–532
44. Nakata K, Fujishima A (2012) TiO_2 photocatalysis: design and applications. J Photochem Photobiol C-Photochem Rev 13(3):169–189
45. Nowotny J, Sorrell CC, Sheppard LR, Bak T (2005) Solar-hydrogen: environmentally safe fuel for the future. Int J Hydrog Energy 30:521–544
46. US Department of Energy (2009) Hydrogen, fuel cells & infrastructure technologies program, Hydrogen production, DOE/GO-102007-2430. p. 3.1-20. http://www.eere.energy.gov/hydrogenandfuelcells/mypp
47. Du P, Schneider J, Li F, Zhao W, Patel U, Castellano F, Eisenberg R (2008) Bi- and terpyridyl platinum(II) chloro complexes: molecular catalysts for the photogeneration of hydrogen from water or simply precursors for colloidal platinum? J Am Chem Soc 130:5056–5058
48. Lei P, Hedlund M, Lomoth R, Rensmo H, Johansson O, Hammarström L (2008) The role of colloid formation in the photoinduced H_2 production with a RuII-PdII supramolecular complex: a study by GC, XPS, and TEM. J Am Chem Soc 130:26–27
49. Streich D, Astuti Y, Orlandi M, Schwartz L, Hammarström L (2010) High-turnover photochemical hydrogen production catalyzed by a model complex of the [FeFe] -hydrogenase active site. Chem Eur J 16:60–63
50. Frey M (2002) Hydrogen-activating enzymes. Chem Bio Chem 3(2–3):153–160
51. Armstrong FA (2004) Hydrogenases: active site puzzles and progress. Curr Opin Chem Biol 8:133–140
52. Evans DJ, Pickett CJ (2003) Chemistry and the hydrogenases. Chem Soc Rev 32:268–275
53. Steinfeld A (2005) Solar thermochemical production of hydrogen-a review. Sol Energy 78:603–615
54. Hollmuller P, Joubert JM, Lachal B, Yvon K (2000) Evaluation of a 5 kWp photovoltaic hydrogen production and storage installation for a residential home in Switzerland. Int J Hydrog Energy 25:97–109
55. Bilgen E (2001) Solar hydrogen from photovoltaic-electrolyzer system. Energy Convers Manag 42:1047–1057
56. Lehman PA, Chamberlin CE, Pauletto G, Rocheleau MA (1997) Operating experience with a photovoltaic-hydrogen energy system. Int J Hydrog Energy 22(5):465–470
57. Rzayeva MP, Salamov OM, Kerimov MK (2001) Modeling to get hydrogen and oxygen by solar water electrolysis. Int J Hydrog Energy 26:195–201
58. Steinfeld A, Meier A (2004) Solar fuelsand materials. In: Encyclopedia of energy, Cleveland (Ed.). Elsevier, Amsterdam, pp 623–637
59. Fletcher EA, Moen RL (1977) Hydrogen and oxygen from water. Science 197:1050–1056
60. Bilgen E (1984) Solar hydrogen production by direct water decomposition process: a preliminary engineering assessment. Int J Hydrog Energy 9:53–58
61. Kogan A (1998) Direct solar thermal splitting of water and onsite separation of the products. II. Experimental feasibility study. Int J Hydrog Energy 23:89–98
62. Ihara S (1980) On the study of hydrogen production from water using solar thermal energy. Int J Hydrog Energy 5:527–534
63. Fletcher EA (1999) Solar thermal and solar quasi-electrolytic processing and separations: zinc from zinc oxide as an example. Ind Eng Chem Res 38:2275–2282
64. Diver RB, Pederson S, Kappauf T, Fletcher EA (1983) Hydrogen and oxygen from water-VI. Quenching the effluent from a solar furnace. Energy 12:947–955

65. Lede J, Villermaux J, Ouzane R, Hossain MA, Ouahes R (1987) Production of hydrogen by simple impingement of a turbulent jet of steam upon a high temperature zirconia surface. Int J Hydrog Energy 12:3–11
66. Olalde G, Gauthier D, Vialaron A (1988) Film boiling around a zirconia target. Application to water thermolysis. Adv Ceram 24:879–883
67. Baykara SZ (2004) Experimental solar water thermolysis. Int J Hydrog Energy 29:1459–1469
68. Baykara SZ (2004) Hydrogen production by direct solar thermal decomposition of water, possibilities for improvement of process efficiency. Int J Hydrog Energy 29:1451–1458
69. Steinfeld A, Brack M, Meier A, Weidenkaff A, Wuillemin D (1998) A solar chemical reactor for the co-production of zinc and synthesis gas. Energy 23:803–814
70. Steinfeld A, Kuhn P, Reller A, Palumbo R, Murray JP, Tamaura Y (1998) Solar-processed metals as clean energy carriers and water-splitters. Int J Hydrog Energy 23:767–774
71. Rodat S, Abanades S, Flamant G (2009) High-temperature solar methane dissociation in a multi tubular cavity-type reactor in the temperature range 1823–2073 K. Energy Fuel 23:2666–2674
72. Abanades S, Flamant G (2008) High-temperature solar chemical reactors for hydrogen production from natural gas cracking. Chem Eng Comm 195:1159–1175. doi:10.1080/00986440801943602
73. Giaconia A, de Falco M, Caputo G, Grena R, Tarquini P, Marrelli L (2008) Solar steam reforming of natural gas for hydrogen production using molten salt heat carriers. Published online in Wiley InterScience (www.interscience.wiley.com). doi: 10.1002/aic.11510
74. Z'Graggena A, Steinfeld A (2008) Hydrogen production by steam-gasification of carbonaceous materials using concentrated solar energy – V. Reactor modeling, optimization, and scale-up. Int J Hydrog Energy 33:5484–5492
75. Piatkowski N, Steinfeld A (2008) Solar-driven coal gasification in a thermally irradiated packed-bed reactor. Energy Fuel 22:2043–2052
76. Dara SS, Singh AK (2008) Basics of engineering chemistry. 2nd revised edition. S. Chand and Company Ltd, New Delhi
77. Hongguang J, Shien S, Wei H, Lin G (2009) Proposal of a novel multifunctional energy system for cogeneration of coke, hydrogen, and power. J Eng Gas Turbine Power 131/052001-1, doi: 10.1115/1.3078791
78. Turner J, Sverdrup G, Mann MK, Maness P-C, Kroposki B, Ghirardi M, Evans RJ, Blake D (2008) Renewable hydrogen production. Int J Energy Res 32:379–407
79. Koroneos C, Dompros A, Roumbas G (2008) Hydrogen production via biomass gasification-a life cycle assessment approach. Chem Eng Proc 47:1261–1268
80. Lv P, Wu C, Ma L, Yuan Z (2008) A study on the economic efficiency of hydrogen production from biomass residues in China. Renew Energy 33:1874–1879
81. Doenitz WR, Schnitedberger R, Steinfield E, Strucker R (1980) Hydrogen production by high temperature electrolysis of water vapor. Int J Hydrog Energy 5:55–63
82. Westinghouse Electric Corporation (1983) Technical progress report no 1: high temperature water vapor electrolysis using solid electrolysis cells, BNL contract no 585847-S, Pittsburgh
83. Lipa MA, Borhan A (1983) High temperature steam electrolysis: technical and economic evaluation of alternative process designs, BNL 51798, Brookhaven National Laboratory, Upton 11973
84. Steinberg M, Cheng HC (1989) Modern and prospective technologies for hydrogen production from fossil fuels. Int J Hydrog Energy 14(11):797–820
85. Reber JF, Meier K (1984) Photochemical production of hydrogen with zinc sulfide suspensions. J Phys Chem 88:5903–5913
86. Bubler N, Reber JF, Meier K (1984) Photochemical production of hydrogen with cadmium sulfide suspensions. J Phys Chem 88:3261–3268
87. Arachchige SM, Brown J, Brewer KJ (2008) Photochemical hydrogen production from water using the new photocatalyst [{(bpy)$_2$Ru(dpp)}$_2$2RhBr$_2$](PF$_6$)$_5$. J Photochem Photobiol A Chem 197:13–17

88. Baniasadi E, Dincer I, Naterer GF (2013) Hybrid photocatalytic water splitting for an expanded range of the solar spectrum with cadmium sulfide and zinc sulfide catalysts. Chem Eng Sci 87:125–139
89. Crabtree RH (2010) Energy production and storage: inorganic chemical strategies for a warming world. Wiley, Chichester
90. Vayssieres L (2010) Solar hydrogen and nanotechnology. Wiley, New York
91. De lasa H, Serrano-Rosales B (2009) Advances in Chemical Engineering: Photocatalytic Technologies, Vol 36 Elsevier, August 2009. ISBN: 978-0-12-374763-1, ISSN: 0065–2377
92. Fujishima A, Honda K (1972) Electrochemical photolysis of water at a semiconductor electrode. Nature 238:37–38
93. Akkerman I, Janssen M, Rocha J, Wijffels RH (2002) Photobiological hydrogen production: photochemical efficiency and bioreactor design. Int J Hydrog Energy 27(11–12):1195–1208
94. Funk JE, Reinstrom RM (1967) Energy requirements in production of hydrogen from water. Ind Eng Chem Proc Des Dev 5(3):336–342
95. Naterer GF, Suppiah S, Stolberg L, Lewis M, Wang Z, Daguppati V, Gabriel K, Dincer I, Rosen MA, Spekkens P, Lvov SN, Fowler F, Tremaine P, Mostagimi J, Easton EB, Trevani L, Rizvi G, Ikeda BM, Kaye MH, Lu L, Pioro I, Smith WR, Seknik E, Jiang J, Avsec J (2010) Canada's program on nuclear hydrogen production and the thermochemical Cu-Cl cycle. Int J Hydrog Energy 35(20):10905–10926
96. Baniasadi E, Dincer I, Naterer GF (2013) Radiation diffraction in a photocatalytic hydrogen production reactor with light scattering and absorption. Int J Hydrog Energy 38 (1):19321–19339
97. Ni M, Leung MKH, Leung DYC, Sumathy K (2007) A review and recent developments in photocatalytic water-splitting using TiO_2 for hydrogen production. Renew Sustain Energy Rev 11(3):401–425
98. McConnell I, Li G, Brudvig GW (2010) Energy conversion in natural and artificial photosynthesis. Chem Biol 17:434–447
99. Barber JM (2009) Photosynthetic energy conversion: natural and artificial. Chem Soc Rev 38:185–196
100. Benniston AC, Harriman A (2008) Artificial photosynthesis. Mater Today 11:26–34
101. Gust D, Kramer D, Moore A, Moore TA, Vermaas W (2008) Engineered and artificial photosynthesis: human ingenuity enters the game. MRS Bull 33:383–387
102. Gust D, Moore TA, Moore AL (2001) Mimicking photosynthetic solar energy transduction. Acc Chem Res 34:40–48
103. Fukuzumi S (2008) Development of bioinspired artificial photosynthetic systems. Phys Chem Chem Phys 10:2283–2297
104. Wenger OS (2009) Long-range electron transfer in artificial systems with d(6) and d(8) metal photosensitizers. Coord Chem Rev 253:1439–1457
105. Durr H, Bossmann S (2001) Ruthenium polypyridine complexes, on the route to biomimetic assemblies as models for the photosynthetic reaction center. Acc Chem Res 34:905–917
106. Wasielewski MR (2009) Self-assembly strategies for integrating light-harvesting and charge separation in artificial photosynthetic systems. Acc Chem Res 42:1910–1921
107. Hasobe T (2010) Supramolecular nano architectures for light energy conversion. Phys Chem Chem Phys 12:44–57
108. Andreiadis ES, Chavarot-Kerlidou M, Fontecave M (2011) Artificial photosynthesis: from molecular catalysts for light-driven water splitting to photoelectrochemical cells. Photochem Photobiol 87(5):946–964
109. Masaoka S, Mukawa Y, Sakai K (2010) Frontier orbital engineering of photo-hydrogen-evolving molecular devices: a clear relationship between the H_2-evolving activity and the energy level of the LUMO. Dalton Trans 39:5868
110. Zhang P, Wang M, Li C, Li X, Dong J, Sun L (2010) Photochemical H(2) production with noble-metal-free molecular devices comprising a porphyrin photosensitizer and a cobaloxime catalyst. Chem Commun 46(46):8806–8808

111. Wang M, Sun L (2010) Hydrogen production by noble-metal-free molecular catalysts and related nanomaterials. Chem Sus Chem 3(5):551–554

112. Fihri A, Artero V, Razavet M, Baffert C, Leibl W, Fontecave M (2008) Cobaloxime-based photocatalytic devices for hydrogen production. Angew Chem Int Ed 47:564–567

113. Fihri A, Artero V, Pereira A, Fontecave M (2008) Efficient H_2-producing photocatalytic systems based on cyclometalated iridium- and tricarbonylrhenium-diimine photosensitizers and cobaloxime catalysts. Dalton Trans 39(25):5567–5569

114. Li C, Wang M, Pan JX, Zhang P, Zhang R, Sun LC (2009) Photochemical hydrogen production catalyzed by polypyridyl ruthenium-cobaloximeheterobinuclear complexes with different bridges. J Organomet Chem 694:2814–2819

115. Brauns E, Jones SW, Clark JA, Molnar SM, Kawanishi Y, Brewer KJ (1997) Electrochemical, spectroscopic and spectroelectrochemical properties of synthetically useful supramolecular light absorbers with mixed-polyazine ligands. Inorg Chem 36:2861–2867

116. Inagaki A, Akita M (2010) Visible-light promoted bimetallic catalysis. Coord Chem Rev 254:1220

117. Kanan MW, Nocera DG (2009) Cobalt-phosphate oxygen-evolving compound. Chem Soc Rev 38:13

118. Tinker LL, McDaniel ND, Bernhard S (2009) Progress towards solar-powered homogeneous water photolysis. J Mater Chem 19:3328

119. Wang M, Na Y, Gorlov M, Sun LC (2009) Light-driven hydrogen production catalysed by transition metal complexes in homogeneous systems. Dalton Trans 33:6458–6467

120. Cline ED, Adamson SE, Bernhard S (2008) Homogeneous catalytic system for photoinduced hydrogen production utilizing iridium and rhodium complexes. Inorg Chem 47:10378

121. Curtin PN, Tinker LL, Burgess CM, Cline ED, Bernhard S (2009) Structure-activity correlations among iridium(III) photosensitizers in a Robust water-reducing system. Inorg Chem 48:10498

122. Du PW, Knowles K, Eisenberg R (2008) A homogeneous system for the photogeneration of hydrogen from water based on a platinum(II) terpyridyl acetylide chromophore and a molecular cobalt catalyst. J Am Chem Soc 130(38):12576–12577

123. Arachchige SM, Shaw R, White TA, Shenoy V, Tsui HM, Brewer KJ (2011) High turnover in a photocatalytic system for water reduction to produce hydrogen using a Ru, Rh, Ru photoinitiated electron collector. Chem Sus Chem 4:1–6

124. Weber MF, Dignam MJJ (1984) Efficiency of splitting water with semiconducting photoelectrode. Electrochem Soc 131(6):1258

125. Zou ZG, Arakawa HJ (2003) Direct water splitting into H_2 and O_2 under visible light irradiation with a new series of mixed oxide semiconductor photo-catalysts. Photochem Photobiol A 158(2–3):145–162

126. Abe R, Sayama K, Sugihara H (2005) Development of new photocatalytic water splitting into H(2) and O(2) using two different semiconductor photocatalysts and a shuttle redox mediator IO(3) (−)/I(−). J Phys Chem B 109(33):16052–16061

127. Kato H, Asakura K (2003) Kudo, highly efficient water splitting into H-2 and O-2 over lanthanum-doped $NaTaO_3$ photocatalysts with high crystallinity and surface nanostructure. A J Am Chem Soc 125(10):3082–3089

128. Reber JF, Meier K (1984) Photochemical production of hydrogen with zinc-sulfide suspensions. J Phys Chem 88(24):5903–5913

129. Maeda K, Teramura K, Lu DL, Takata T, Saito N, Inoue Y, Domen KJ (2007) Characterization of ruthenium oxide nanocluster as a cocatalyst with (Gal(1-x)Zn(x))(N1-xOx) for photocatalytic overall water splitting. J Phys Chem B 110(28):13753–13758

130. Maeda K, Teramura K, Lu DL, Takata T, Saito N, Inoue Y, Domen K (2007) Photocatalyst releasing hydrogen from water – enhancing catalytic performance holds promise for hydrogen production by water splitting in sunlight. Nature 440(7082):295–295

131. Nowotny J, Sorrell CC, Bak T, Sheppard LR (2005) Solar-hydrogen: unresolved problems in solid-state science. Sol Energy 78(5):593–602

132. Maeda K, Domen K (2007) New non-oxide photocatalysts designed for overall water splitting under visible light. J Phys Chem C 111(22):7851–7861

133. Chen X, Mao S (2007) Titanium dioxide nanomaterials: synthesis, properties, modifications, and applications. Chem Rev 107(7):2891–2959

134. Kitano M, Takeuchi M, Matsuoka M, Thomas JM, Anpo M (2007) Photocatalytic water splitting using Pt-loaded visible light-responsive TiO(2) thin film photocatalysts. Catal Today 120(2):133

135. Selli E, Chiarello GL, Quartarone E, Mustarelli P, Rossetti I, Forni L (2007) A photocatalytic water splitting device for separate hydrogen and oxygen evolution. Chem Commun 47:5022–5024

136. Sun LC, Li L, Duan L, Xu Y, Gorlov M (2010) A photoelectrochemical device for visible light driven water splitting by a molecular ruthenium catalyst assembled on dye-sensitized nanostructured TiO(2). Chem Commun 46(39):7307–7309

137. Baniasadi E, Dincer I, Naterer GF (2012) Performance analysis of a water splitting reactor with hybrid photochemical conversion of solar energy. Int J Hydrog Energy 37:7464–7472

138. Baniasadi E, Dincer I, Naterer GF (2012) Radiative heat transfer and catalyst performance in a large-scale continuous flow photoreactor for hydrogen production. Chem Eng Sci 84:638–645

139. Kelly NA, Gibson TL (2008) Solar energy concentrating reactors for hydrogen production by photoelectrochemical water splitting. Int J Hydrog Energy 33(22):6420–6431

140. Fujishima A, Honda K (1972) Electrochemical photolysis of water at a semiconductor electrode. Nature 238:37–38

141. Nowotny J, Sorrell CC, Bak T, Sheppard LR (2005) Solar-hydrogen: unresolved problems in solid-state science. Sol Energy 78:593–602

142. Photoelectrochemical Cell (PEC), Lawrence Berkeley National Laboratory, Materials Sciences Division, Solar Energy Materials Research Group. http://emat-solar.lbl.gov/research/PEC.html. Accessed in Apr 2010

143. Yamada Y, Matsuki N, Ohmori T, Mametsuka H, Kondo M, Matsuda A (2003) One chip photovoltaic water electrolysis device. Int J Hydrog Energy 28:1167–1169

144. Dincer I (2002) Technical, environmental and exergetic aspects of hydrogen energy systems. Int J Hydrog Energy 27(3):265–285

145. Licht S, Wang B, Mukerji S, Soga T, Umeno M, Tributsch H (2001) Over 18 % solar energy conversion to generation of hydrogen fuel; theory and experiment for efficient solar water splitting. Int J Hydrog Energy 26:653–659

146. Murphy OJ, Bockris JO'M (1984) Photovoltaic electrolysis: hydrogen and electricity from water and light. Int J Hydrog Energy 9(7):557–561

147. Kotay SM, Das D (2008) Biohydrogen as a renewable energy resource – prospects and potentials. Int J Hydrog Energy 33:258–263

148. Luzzi A, Bonadio L, McCann M, editors (2004) In pursuit of the future – 25 years of IEA research towards the realisation of hydrogen energy systems. International Energy Agency – Hydrogen Implementing Agreement 2004

149. Benemann JR (1997) Feasibility analysis of photobiological hydrogen production. Int J Hydrog Energy 22:979–987

150. Sturzenegger M, Nu¨esch P (1999) Efficiency analysis for a manganese-oxide-based thermochemical cycle. Energy 24(1999):959–970

151. Tyagi SK, Wang S, Singhal MK, Kaushik SC, Park SR (2007) Exergy analysis and parametric study of concentrating type solar collectors. Int J Therma Sci 46:1304–1310

152. Crouse WH, Anglin DL (1985) Automotive mechanics, 3rd edn. Tata McGraw-Hill Publishing Company Limited, New York, pp 375–385

153. Evans A, Strezov V, Evans TJ (2008) Assessment of sustainability indicators for renewable energy technologies. Renew Sustain Energy Rev 13: 1082–1088, www.elsevier.com/locate/rser

154. Rosen MA, Dincer I, Kanoglu M (2008) Role of exergy in increasing efficiency and sustainability and reducing environmental impact. Energy Policy 36:128–137

155. Connelly L, Koshland CP (1997) Two aspects of consumption: using an exergy-based measure of degradation to advance the theory and implementation of industrial ecology. Resour Conserv Recycl 19:199–217

156. Collings AF, Critchley C (2005) Artificial photosynthesis from basic biology to industrial application. WILEY-VCH Verlag GmbH & Co. KGaA, Weinheim

157. Toolbox-The Engineering Toolbox (2010) http://www.engineeringtoolbox.com/overall-heat-transfer-coefficient-d_434.html. Accessed May 2010

158. Joshi AS, Dincer I, Reddy BV (2010) Solar hydrogen production: a comparative performance assessment. Int J Hydrog Energy 36, doi: 10.1016/j.ijhydene.2010.11.122

159. Dincer I (2007) Environmental and sustainability aspects of hydrogen and fuel cell systems. Int J Energy Res 31:29–55

160. Dincer I, Balta MT (2011) Potential thermochemical and hybrid cycles for nuclear-based hydrogen production. Int J Energy Res 35:123–127

161. Naterer GF, Suppiah S, Stolberg L, Lewis M, Ferrandon M, Wang Z, Dincer I, Gabriel K, Rosen MA, Secnik E, Easton EB, Trevani L, Pioro I, Tremaine P, Lvov S, Jiang J, Rizvi G, Ikeda BM, Lu L, Kaye M, Smith WR, Mostaghimi J, Spekkens P, Fowler M, Avsec J (2011) Clean hydrogen production with the Cu-Cl cycle – progress of international consortium, II: simulations, thermochemical data and materials. Int J Hydrog Energy 36:15486–15501

162. Ozbilen A, Dincer I, Naterer GF, Aydin M (2012) Role of hydrogen storage in renewable energy management for Ontario. Int J Hydrog Energy 37:7343–7354

163. Boehm R, Chen Y, Earl B, Hsieh S, Moujaes S (2003) H₂ technology survey. UNLV program, University of Nevada Las Vegas, Center for Energy Research, November 25. www.unlv.edu

164. Ozbilen A, Aydin M, Dincer I, Rosen MA (2013) Life cycle assessment of nuclear-based hydrogen production via a copper-chlorine cycle: A neural network approach. Int J Hydrog Energy 38:6314–6322

165. Ozbilen A, Dincer I, Rosen MA (2012) Life cycle assessment of hydrogen production via thermochemical water splitting using multi-step Cu-Cl cycles. J Clean Prod 33:202–216

166. Lenzen M (1999) Greenhouse gas analysis of solar-thermal electricity generation. Sol Energy 65:353–368

167. Parry ML, Canziani OF, Palutikof JP, van der Linden PJ, Hanson CE (2007) Contribution of working group II to the fourth assessment report of the intergovernmental panel on climate change, 2007. Cambridge University Press, Cambridge, UK\New York

168. Ozbilen A, Dincer I, Rosen MA (2013) Comparative environmental impact and efficiency assessment of selected hydrogen production methods. Environ Impact Asses 42:1–9

169. Giaconia A, Sau S, Felici C, Tarquini P, Karaginnakis G (2011) Hydrogen production via sulfur-based thermochemical cycles: part 2: performance evaluation of Fe2O3-based catalysts for the sulfuric acid decomposition step. Int J Hydrog Energy 36:6496–6509

170. Aghahosseinin S, Dincer I, Naterer G (2011) Integrated gasification and Cu-Cl cycle for trigeneration of hydrogen, steam and electricity. Int J Hydrog Energy 36:2845–2854

171. Naterer GF, Suppiah S, Stilberg L, Lewis M et al (2011) Clean hydrogen production with the Cu-Cl cycle – progress of international consortium, II: simulations, thermochemical data and materials. Int J Hydrog Energy 36:15486–501

172. Naterer GF, Suppiah S, Stilberg L, Lewis M et al (2011) Clean hydrogen production with the Cu-Cl cycle – progress of international consortium I: experimental unit operations. Int J Hydrog Energy 36:15472–15485

173. Ferrandon MS, Lewis MA, Tatterson DF, Gross A et al (2010) Hydrogen production by the Cu–Cl thermochemical cycle: investigation of the key step of hydrolysing $CuCl_2$ to Cu_2OCl_2 and HCl using a spray reactor. Int J Hydrog Energy 35:992–1000

174. Lewis MA, Masin JG, Vilim RB, Serban M (2010) Development of the low temperature Cu-Cl cycle. Proceedings 2005 International Congress on Advances in Nuclear Power Plants, May 15–19, 2005. Seoul, Korea: American Nuclear Society; May 2005

175. Zamfirescu C, Dincer I, Naterer G (2010) Thermophysical properties of copper compounds in copper–chlorine thermochemical water splitting cycles. Int J Hydrog Energy 35:4839–4852
176. Thomas LG, Nelson AK (2010) Predicting efficiency of solar powered hydrogen generation using photo voltaic electrolysis devices. Int J Hydrog Energy 35:900–911
177. Koroneos C, Dompros A, Roumbas G, Moussiopoulos N (2004) Life cycle assessment of hydrogen fuel production processes. Int J Hydrog Energy 29:1443–1450
178. Ratlamwala TAH, Gadalla MA, Dincer I (2011) Performance assessment of an integrated PV/T and triple effect cooling system for hydrogen and cooling production. Int J Hydrog Energy 36:11282–11291
179. Ratlamwala TAH, Dincer I (2012) Energy and exergy analyses of a Cu–Cl cycle based integrated system for hydrogen production. Chem Eng Sci 84:564–573